KT-561-853

Practice Papers for SQA Exams

Higher

Mathematics

ISBN 978-1-84372-783-5

Published by
Leckie & Leckie Ltd
An imprint of HarperCollins*Publishers*
Westerhill Road, Bishopbriggs, Glasgow, G64 2QT
T: 0844 576 8126   F: 0844 576 8131
leckieandleckie@harpercollins.co.uk   www.leckieandleckie.co.uk

A CIP Catalogue record for this book is available from the British Library.

Questions and answers in this book do not emanate from SQA. All of our entirely new and original Practice Papers have been written by experienced authors working directly for the publisher.

Mixed Sources
Product group from well-managed
forests and other controlled sources
www.fsc.org Cert no. SW-COC-001806
© 1996 Forest Stewardship Council

FSC is a non-profit international organisation established to promote the responsible management of the world's forests. Products carrying the FSC label are independently certified to assure consumers that they come from forests that are managed to meet the social, economic and ecological needs of present and future generations.

Find out more about HarperCollins
and the environment at
www.harpercollins.co.uk/green

Introduction

Layout of the Book

This book contains practice exam papers, which mirror the actual SQA exam as much as possible. The layout, paper colour and question level are all similar to the actual exam that you will sit, so that you are familiar with what the exam paper will look like.

The solutions section is at the back of the book. The full worked solution is given to each question so that you can see how the right answer has been arrived at. The solutions are accompanied by a commentary which includes further explanations and advice. There is also an indication of how the marks are allocated and, where relevant, what the examiners will be looking for. Reference is made at times to the relevant sections in Leckie and Leckie's book 'Higher Maths Revision Notes'.

Revision advice is provided in this introductory section of the book, so please read on!

How To Use This Book

The Practice Papers can be used in two main ways:

1. You can complete an entire practice paper as preparation for the final exam. If you would like to use the book in this way, you might want to complete each practice paper under exam-style conditions by setting yourself a time for each paper and answering it as well as possible without using any references or notes. Alternatively, you can answer the practice paper questions as a revision exercise, using your notes to produce a model answer. Your teacher may mark these for you.

2. You can use the Topic Index at the front of this book to find all the questions within the book that deal with a specific topic. This allows you to focus specifically on areas that you particularly want to revise or, if you are mid-way through your course, it lets you practise answering exam-style questions for just those topics that you have studied.

Revision Advice

Work out a revision timetable for each week's work in advance – remember to cover all of your subjects and to leave time for homework and breaks. For example:

Day	6pm–6.45pm	7pm–8pm	8.15pm–9pm	9.15pm–10pm
Monday	Homework	Homework	English Revision	Chemistry Revision
Tuesday	Maths Revision	Physics Revision	Homework	Free
Wednesday	Geography Revision	Modern Studies Revision	English Revision	French Revision
Thursday	Homework	Maths Revision	Chemistry Revision	Free
Friday	Geography Revision	French Revision	Free	Free
Saturday	Free	Free	Free	Free
Sunday	Modern Studies Revision	Maths Revision	Modern Studies Revision	Homework

Make sure that you have at least one evening free a week to relax, socialise and re-charge your batteries. It also gives your brain a chance to process the information that you have been feeding it all week.

Arrange your study time into one hour or 30 minute sessions, with a break between sessions, e.g. 6pm – 7pm, 7.15pm – 7.45pm, 8pm-9pm. Try to start studying as early as possible in the evening when your brain is still alert and be aware that the longer you put off starting, the harder it will be to start!

Study a different subject in each session, except for the day before an exam.

Do something different during your breaks between study sessions – have a cup of tea, or listen to some music. Don't let your 15 minutes expand into 20 or 25 minutes though!

Have your class notes and any textbooks available for your revision to hand as well as plenty of blank paper, a pen, etc. You should take note of any topic area that you are having particular difficulty with, as and when the difficulty arises. Revisit that question later having revised that topic area by attempting some further questions from the exercises in your textbook.

Revising for a Maths exam is different from revising for some of your other subjects. Revision is only effective if you are trying to solve problems. You may like to make a list of 'Key Questions' with the dates of your various attempts (successful or not!). These should be questions that you have had real difficulty with.

Key Question	1st Attempt		2nd Attempt		3rd Attempt	
Textbook P56 Q3a	18/2/10	X	21/2/10	✓	28/2/10	✓
Practice Exam A Paper1 Q5	25/2/10	X	28/2/10	X	3/3/10	
2008 SQA Paper, Paper2 Q4c	27/2/10	X	2/3/10			

The method for working this list is as follows:

1. Any attempt at a question should be dated.

2. A tick or cross should be entered to mark the success or failure of each attempt.

3. A date for your next attempt at that question should be entered:
   for an unsuccessful attempt – 3 days later
   for a successful attempt – 1 week later

4. After two successful attempts remove that question from the list
   (you can assume the question has been learnt!)

Using 'The List' method for revising for your Maths exam ensures that your revision is focused on the difficulties you have had and that you are actively trying to overcome these difficulties.

Finally, forget or ignore all or some of the advice in this section if you are happy with your present way of studying. Everyone revises differently, so find a way that works for you!

Transfer Your Knowledge

As well as using your class notes and textbooks to revise, these practice papers will also be a useful revision tool as they will help you to get used to answering exam style questions. As you work through the questions, you may find an example that you haven't come across before. Don't worry! There may be several reasons for this. This question may be on a topic that you have not yet covered in class. Check with your teacher to find out if this is the case. Or you may find that the wording or the context of the question is unfamiliar. This often happens with reasoning questions in the Maths exam. Once you have familiarised yourself with the worked solutions you will find that, in most cases, the question is using mathematical techniques with which you are familiar. In either case you should revisit that question later to check that you can successfully solve it.

Trigger Words

In the practice papers and in the exam itself, a number of 'trigger words' will be used in the questions. These trigger words should help you identify a process or a technique that is expected in your solution to that part of the question. If you familiarise yourself with these trigger words, it will help you to structure your solutions more effectively.

Trigger Word	Meaning / Explanation
Evaluate	Carry out a calculation to give an answer that is a value.
Hence	You must use the result of the previous part of the question to complete your solution. No marks will be given if you use an alternative method that does not use the previous answer.

Simplify	This means different things in different contexts:
	Surds: reduce the number under the root sign to the smallest possible by removing square factors.
	Fractions: one fraction, cancelled down, is expected.
	Algebraic expressions: get rid of brackets and gather all like terms together.
Give your answer to …	This is an instruction for the accuracy of your final answer. These instructions must be followed or you will lose a mark.
Algebraically	The method you use must involve algebra, e.g. you must solve an equation or simplify an algebraic equation. It is usually stated to avoid trial-and-improvement methods or reading answers from your calculator.
Justify your answer	This is a request for you to clearly indicate your reasoning. Will the examiner know how your answer was obtained?
Show all your working	Marks will be allocated for the individual steps in your working. Steps missed out may lose you marks.

In the Exam

Watch your time and pace yourself carefully. You will find some questions harder than others. Try not to get stuck on one question as you may later run out of time. Instead, move on and then return to a difficult question later. Remember also that if you have spare time towards the end of your exam, use this time to check through your solutions. Mistakes often are discovered in this checking process and can be corrected.

Become familiar with the exam instructions. The practice papers in this book have exam instructions at the front of each exam. Also remember that there is a formulae list to consult. You will find this at the front of your exam paper. However, even though these formulae are given to you, it is important that you learn them so that they are familiar to you. If you are continuing with Mathematics next session it will be assumed that these formulae are known in next year's exam!

Read the question thoroughly before you begin to answer it – make sure you know exactly what the question is asking you to do. If the question is in sections, e.g. 15a, 15b, 15c, etc, then it is often the case that answers obtained in the earlier sections are used in the later sections of that question.

When you have completed your solution read it over again. Is your reasoning clear? Will the examiner understand how you arrived at your answer? If in doubt then fill in more details.

If you change your mind or think that your solution is wrong, don't score it out unless you have another solution to replace it with. Solutions that are not correct can often gain some of the marks available. Do not miss working out. Showing step-by-step working will help you gain maximum marks even if there is a mistake in the working.

Use these resources constructively by reworking questions later that you found difficult or impossible first time round. Remember: success in a Maths exam will only come from actively trying to solve lots of questions and only consulting notes when you are stuck. Reading notes alone is not a good way to revise for your Maths exam. Always be active, always solve problems.

Good luck!

Topic Index

Practice Papers for SQA Exams: Higher Mathematics

Topic	A Paper 1	A Paper 2	B Paper 1	B Paper 2	C Paper 1	C Paper 2	Knowledge for Prelim: Have difficulty	Still needs work	OK	Knowledge for SQA Exam: Have difficulty	Still needs work	OK
Unit 1												
• The Straight Line	8	2	3,9	2	2,6,14							
• Functions and Graphs	3,4,10,13		7,17,19,23	7	9,11							
• Trig–Basic Facts	1,19			5	23							
• Intro to Differentiation	2,21	1,6	21,22	7	13,22	4,7,8						
• Recurrence Relations	9,23		2,8		1,4							
Unit 2												
• Polynomials	22		22		3,8							
• Quadratic Theory	11		1,15		19	3						
• Intro to Integration	6,15	6		6	25	9						
• Further Trig	5,24	7	5,10	3	5,23							
• Circles	7,12	1	4	1		6						
Unit 3												
• Vectors	14,18	3	6,13,14,18	4	7,10,12,17	2						
• Further Diff & Int	17		11,12		16,18,20							
• Log & Exp Functions	16,20	5	16,20,23	5	15,24	5						
• Wave Function		4	4		21	1						

Practice Exam A

Mathematics Higher

Practice Papers
for SQA Exams

**Exam A
Higher
Paper 1
Non-calculator**

You are allowed 1 hour, 30 minutes to complete this paper.

You must not use a calculator.

Full marks will only be awarded where your answers include relevant working.

You will not receive any marks for answers derived from scale drawings.

Leckie × Leckie
Scotland's leading educational publishers

FORMULAE LIST

Trigonometric formulae

$$\sin (A \pm B) = \sin A \cos B \pm \cos A \sin B$$
$$\cos (A \pm B) = \cos A \cos B \mp \sin A \sin B$$
$$\sin 2A = 2\sin A \cos A$$
$$\cos 2A = \cos^2 A - \sin^2 A$$
$$= 2\cos^2 A - 1$$
$$= 1 - 2\sin^2 A$$

Circle

The equation $x^2 + y^2 + 2nx + 2py + c = 0$ represents a circle centre $(-n, -p)$ and radius $\sqrt{n^2 + p^2 - c}$.

The equation $(x - a)^2 + (y - b)^2 = r^2$ represents a circle centre (a, b) and radius r.

Table of standard integrals

$f(x)$	$\int f(x)dx$
$\sin ax$	$-\dfrac{1}{a}\cos ax + C$
$\cos ax$	$\dfrac{1}{a}\sin ax + C$

Table of standard derivatives

$f(x)$	$f'(x)$
$\sin ax$	$a \cos ax$
$\cos ax$	$-a \sin ax$

Scalar Product $a.b = |a||b| \cos \theta$, where θ is the angle between a and b

or $a.b = a_1b_1 + a_2b_2 + a_3b_3$ where $a = \begin{pmatrix} a_1 \\ a_2 \\ a_3 \end{pmatrix}$ and $b = \begin{pmatrix} b_1 \\ b_2 \\ b_3 \end{pmatrix}$.

SECTION A

1. The diagram shows a right-angled triangle with sides 2, $2\sqrt{2}$ and $2\sqrt{3}$. What is the value of $\sin 2y°$?

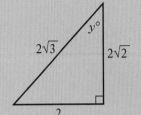

   A $\dfrac{2}{\sqrt{3}}$

   B $\dfrac{1}{3}$

   C $\dfrac{2\sqrt{2}}{3}$

   D $2\sqrt{3}$

2. If $y = \dfrac{x^4 + 1}{x}$ what is $\dfrac{dy}{dx}$?

   A $4x + 1$

   B $x^4 + 1 + x^{-1}$

   C $4x^3 + 1$

   D $3x^2 - \dfrac{1}{x^2}$

3. Which of the following describes the stationary point on the curve with equation $y = 2(x + 2)^3 - 1$?

   A minimum at $(2, -1)$

   B maximum at $(2, -1)$

   C minimum at $(-2, -1)$

   D maximum at $(-2, -1)$

4. Functions f and g are given by $f(x) = \sqrt{x}$ and $g(x) = 3 - x$ for $x \geq 0$.
Which of the following is an expression for $f(g(x))$?

   A $3\sqrt{x} - x\sqrt{x}$

   B $\sqrt{3 - x}$

   C $\sqrt{x} + 3 - x$

   D $3 - x\sqrt{x}$

5. p and q are angles as shown in the diagram
What is the value of $\cos(p + q)°$?

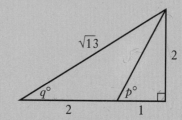

   A $\dfrac{8}{\sqrt{65}}$

   B $-\dfrac{1}{\sqrt{65}}$

   C $\dfrac{1}{\sqrt{5}} + \dfrac{3}{\sqrt{13}}$

   D $\dfrac{1}{\sqrt{5}} - \dfrac{3}{\sqrt{13}}$

6. Find $\displaystyle\int \dfrac{3}{\sqrt{x}}\,dx$

   A $6\sqrt{x} + c$

   B $-\dfrac{3}{x} + c$

   C $\dfrac{3}{2\sqrt{x}} + c$

   D $\dfrac{3}{x^2} + c$

7. A circle has equation $x^2 + y^2 - 2x + 6y - 1 = 0$.
   What is the radius of this circle?

   A $\sqrt{3}$

   B $\sqrt{11}$

   C $\sqrt{37}$

   D $\sqrt{39}$

8. What is the distance PQ where P is the point $(-1, -3, 5)$ and Q is the point $(0, 5, -2)$?

   A $\sqrt{12}$

   B $\sqrt{14}$

   C $\sqrt{64}$

   D $\sqrt{114}$

9. A sequence is defined by the recurrence relation

   $$u_{n+1} = u_n^2 - 1, \quad u_0 = -2$$

   What is the value of u_2?

   A -26

   B -11

   C 8

   D 64

10. The diagram shows a sketch of the graph with equation $y = f(x)$. Which of the diagrams below shows a sketch of $y = -f(x - 4)$?

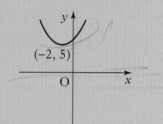

A

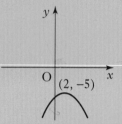

B

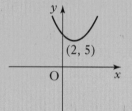

C

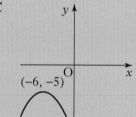

D

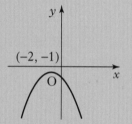

11. The equation $3x^2 + x + m = 0$ has equal roots. What is the value of m?

A $-\dfrac{1}{3}$

B $-\dfrac{1}{12}$

C $\dfrac{1}{12}$

D   12

12. The point A(2, 3) lies on the circle with equation $x^2 + y^2 + 2x - 4y - 5 = 0$. What is the gradient of the tangent at A?

A -10

B -3

C 0

D $\dfrac{1}{3}$

13. $7 - 8x - x^2$ is expressed in the form $a - (x + b)^2$. What is the value of a?

A -23

B -9

C 9

D 23

14. The points A(10, -1, 3), B(6, 1, 1) and C (4, 2, t) are collinear as shown in the diagram. What is the value of t?

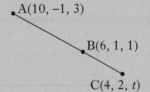

A 0

B 2

C 4

D 6

15. For a curve $y = f(x)$ it is known that $\dfrac{dy}{dx} = 4x^3 - x^2 - 1$ and that it passes through the origin. What is the equation of the curve?

A $y = x^4 - \dfrac{1}{3}x^3 - x$

B $y = 12x^4 - 2x^3$

C $y = 12x^2 - 2x$

D $y = 4x^4 - x^3 - x + 1$

16. If $\log_2 9 = 3 - \log_2 x$, what is the value of x?

A $\dfrac{3}{81}$

B $\dfrac{8}{9}$

C   1

D   6

17. What is the value of $\displaystyle\int_0^{\pi/3} \cos\dfrac{1}{2}x\ dx$?

A $-\sqrt{3}$

B $-\dfrac{1}{2}$

C   1

D $\sqrt{3}$

18. The vectors p, q and r are represented by the sides of an equilateral triangle as shown in the diagram.

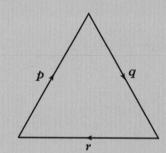

Here are two statements about these vectors:

    (1) $q.(p - r) = 0$

    (2) $q.(p + r) = 0$

Which of the following is true?

A   Neither statement is true.

B   Only statement (1) is correct.

C   Only statement (2) is correct.

D   Both statements are correct.

19. What is the equation of the graph shown in the diagram?

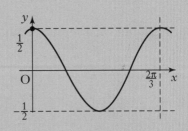

A $y = \dfrac{1}{2}\cos\dfrac{1}{3}x$

B $y = \cos 3x + \dfrac{1}{2}$

C $y = \dfrac{2}{3}\cos\dfrac{1}{2}x$

D $y = \dfrac{1}{2}\cos 3x$

20. If $\dfrac{\log_k 4}{\log_e 2} = 2e^0$, what is the value of k?

A $k = 1$

B $k = \sqrt{2}$

C $k = \sqrt[4]{2e^2}$

D $k = e$

[End of section A]

SECTION B

21. (a) Find the stationary points on the curve with equation $y = x^3 - 3x^2 + 4$ and justify their nature.

    7

    (b) (i) Show that $(x + 1)(x - 2)^2 = x^3 - 3x^2 + 4$

      (ii) Hence sketch the graph of $y = x^3 - 3x^2 + 4$

    4

22. Two cubic graphs, $y = f(x)$ and $y = g(x)$, where $f(x) = 2x^3 + 3x + 12$ and $g(x) = 2 + 16x^2 - x^3$, are shown in the diagram.

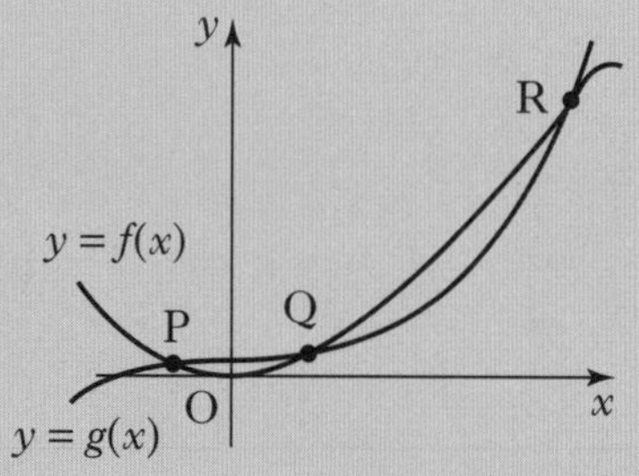

Determine the x–coordinates of each of P, Q and R, the three points of intersection of the two graphs.

    8

23. The islanders living in Tarbert on the island of Harris are planning to build a new sewage processing plant. Central to the plant is the seepage pit which allows most of the week's sewage to seep harmlessly through the soil and drain away. Sewage is pumped into the pit at the start of each week. There are two possible sites with the following specifications:

	Seepage Rate	Pumping capacity
Upland Site:	65% of 1 week's sewage	2000 litres at start of week
Lowland Site:	75% of 1 week's sewage	2500 litres at start of week

    (a) Write down a recurrence relation for each site using u_n to represent the amount of litres of sewage stored at the Upland site immediately after pumping at the start of the nth week and let v_n be the equivalent volume at the Lowland site. Clearly label each relation with the site name.

    2

    (b) The size of the storage tank at each site is determined by the maximum volume of sewage that will remain at the site in the long term. Which site requires the smaller tank in the long term?

    4

24. Solve the equation $\cos \theta(\cos \theta - 1) = \sin^2 \theta$ for $\pi < \theta < 2\pi$

    5

[End of section B]

[End of question paper]

Mathematics   Higher

Practice Papers
for SQA Exams

Exam A
Higher
Paper 2

You are allowed 1 hour, 10 minutes to complete this paper.

You may use a calculator.

Full marks will only be awarded where your answer includes relevant working.

You will not receive any marks for answers derived from scale drawings.

Scotland's leading educational publishers

FORMULAE LIST

Trigonometric formulae

$$\sin (A \pm B) = \sin A \cos B \pm \cos A \sin B$$
$$\cos (A \pm B) = \cos A \cos B \mp \sin A \sin B$$
$$\sin 2A = 2\sin A \cos A$$
$$\cos 2A = \cos^2 A - \sin^2 A$$
$$= 2\cos^2 A - 1$$
$$= 1 - 2\sin^2 A$$

Circle

The equation $x^2 + y^2 + 2nx + 2py + c = 0$ represents a circle centre $(-n, -p)$ and radius $\sqrt{n^2 + p^2 - c}$.

The equation $(x - a)^2 + (y - b)^2 = r^2$ represents a circle centre (a, b) and radius r.

Table of standard integrals

$f(x)$	$\int f(x)dx$
$\sin ax$	$-\dfrac{1}{a}\cos ax + C$
$\cos ax$	$\dfrac{1}{a}\sin ax + C$

Table of standard derivatives

$f(x)$	$f'(x)$
$\sin ax$	$a \cos ax$
$\cos ax$	$-a \sin ax$

Scalar Product $a.b = |a||b| \cos \theta$, where θ is the angle between a and b

or $a.b = a_1 b_1 + a_2 b_2 + a_3 b_3$ where $a = \begin{pmatrix} a_1 \\ a_2 \\ a_3 \end{pmatrix}$ and $b = \begin{pmatrix} b_1 \\ b_2 \\ b_3 \end{pmatrix}$.

1. The diagram shows a cubic curve with equation $y = x^2 - \dfrac{1}{3}x^3$.

A tangent PQ to the curve has point of contact M(3, 0).

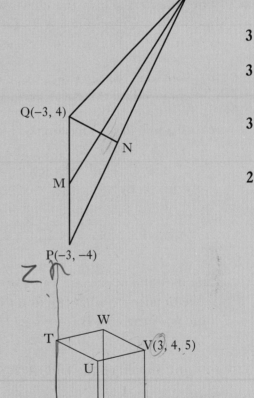

Marks

(a) Find the equation of PQ.　　　　　4

A circle has equation $x^2 + y^2 - 4x - 26y + 163 = 0$

(b) Show that PQ is also a tangent to this circle and find the coordinates of the point of contact N.　　　6

(c) Find the ratio in which the y-axis cuts the line MN.　　　3

2. Triangle PQR has coordinates P(−3, −4), Q(−3, 4) and R(5, 12).

(a) Find the equation of the median MR.　　　3

(b) Find the equation of the altitude NQ.　　　3

(c) Median MR and altitude NQ intersect at point S. Find the coordinates of S.　　　3

(d) The point T(2, 9) lies on QR. Show that ST is parallel to PR.　　　2

3.

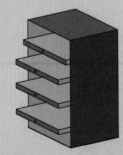

This set of drawers is being 'modelled' on a computer software design package as a cuboid as shown. The edges of the cuboid are parallel to the x, y and z-axes. Three of the vertices are P(−1,−1,−1), S(−1,4,−1) and V(3,4,5).

(a) Write down the lengths of PQ, QR and RV.　　　1

(b) Write down the components of $\overrightarrow{VS}$ and $\overrightarrow{VP}$ and hence calculate the size of angle PVS.　　　7

Marks

4. (a) Express $3\cos x° - 2\sin x°$ in the form $k\cos(x + a)°$
where $k > 0$ and $0 \le a \le 90$.   **4**

(b) Hence solve the equation $3\cos x° - 2\sin x° = 2$ for $0 < x < 360$.   **3**

5. Atmospheric pressure decreases exponentially as you rise above sea-level. It is known that the atmospheric pressure, $P(h)$, at a height h kilometres above sea level is given by $P(h) = P_0 e^{-kh}$ where P_0 is the pressure at sea-level ($h = 0$).

(a) Given that at a height of $4·95$ km the atmospheric pressure is half that at sea-level, calculate the value of k correct to 4 decimal places.   **3**

(b) Mount Everest is 8850 metres high. What is the percentage decrease in air pressure at the top of Mount Everest compared to the pressure at sea-level?   **2**

6. The diagram shows the curve with equation $y = 6 + 4x - x^2$ and the straight line with equation $y = x + 2$. The line intersects the curve at points S and T as shown.

(a) Calculate B, the exact value of the area enclosed by the curve and the line.   **7**

(b) A point $P(x, y)$ lies on the curve between S and T and it is known that the area, A, of triangle PST (shaded in the diagram) is given by

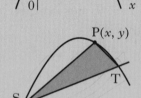

$$A(x) = -\frac{5}{2}x^2 + \frac{15}{2}x + 10$$

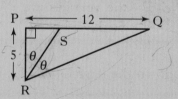

Calculate the maximum value of this area and hence determine what fraction this maximum value is of the area B from part (a).   **4**

7. In right-angled triangle PQR, RS is the bisector of angle PRQ. PR = 5 units and PQ = 12 units. Show that the exact value of $\cos \theta$ is $\dfrac{3\sqrt{3}}{13}$ **5**

[End of question paper]

Practice Exam B

Mathematics　　　　Higher

Practice Papers
for SQA Exams

Exam B
Higher
Paper 1
Non-calculator

You are allowed 1 hour, 30 minutes to complete this paper.

You must not use a calculator.

Full marks will only be awarded where your answer includes relevant working.

You will not receive any marks for answers derived from scale drawings.

Leckie ✕ Leckie
Scotland's leading educational publishers

FORMULAE LIST

Trigonometric formulae

$$\sin (A \pm B) = \sin A \cos B \pm \cos A \sin B$$
$$\cos (A \pm B) = \cos A \cos B \mp \sin A \sin B$$
$$\sin 2A = 2\sin A \cos A$$
$$\cos 2A = \cos^2 A - \sin^2 A$$
$$= 2\cos^2 A - 1$$
$$= 1 - 2\sin^2 A$$

Circle

The equation $x^2 + y^2 + 2nx + 2py + c = 0$ represents a circle centre $(-n, -p)$ and radius $\sqrt{n^2 + p^2 - c}$.

The equation $(x - a)^2 + (y - b)^2 = r^2$ represents a circle centre (a, b) and radius r.

Table of standard integrals

$f(x)$	$\int f(x)dx$
$\sin ax$	$-\dfrac{1}{a}\cos ax + C$
$\cos ax$	$\dfrac{1}{a}\sin ax + C$

Table of standard derivatives

$f(x)$	$f'(x)$
$\sin ax$	$a \cos ax$
$\cos ax$	$-a \sin ax$

Scalar Product $a.b = |a||b| \cos \theta$, where θ is the angle between a and b

or $a.b = a_1b_1 + a_2b_2 + a_3b_3$ where $a = \begin{pmatrix} a_1 \\ a_2 \\ a_3 \end{pmatrix}$ and $b = \begin{pmatrix} b_1 \\ b_2 \\ b_3 \end{pmatrix}$.

SECTION A

1. Here are two statements about the roots of equation $x^2 - x - 2 = 0$

    (1) The roots are rational;

    (2) The roots are real.

    Which of the following is true?

    A   Neither statement is correct.

    B   Only statement (1) is correct.

    C   Only statement (2) is correct.

    D   Both statements are correct.

2. A sequence is defined by the recurrence relation

    $$u_{n+1} = 0{\cdot}8\, u_n + 3 \text{ with } u_0 = 5$$

    What is the value of u_2?

    A $6{\cdot}8$

    B $8{\cdot}6$

    C $19{\cdot}0$

    D $35{\cdot}0$

3. A line AB makes an angle of $50°$ with the positive direction of the y-axis as shown in the diagram.

    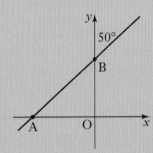

    What is the gradient of line AB?

    A $\tan 130°$

    B $-\dfrac{1}{\tan 50°}$

    C $\tan 50°$

    D $\tan 40°$

4. The line $y = 5$ is a tangent to a circle with centre C(0, 3) as shown in the diagram. What is the equation of the circle?

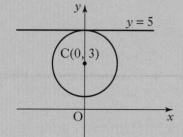

A $\quad x^2 + (y - 3)^2 = 4$

B $\quad x^2 + (y - 3)^2 = 9$

C $\quad x^2 + (y + 3)^2 = 4$

D $\quad (x - 3)^2 + y^2 = 9$

5. Given that $\tan p° = \dfrac{1}{2}$ with $0 \leq p < 90$, which of the following is an expression for $\cos (p - q)°$?

A $\quad \dfrac{2}{\sqrt{5}} - \cos q°$

B $\quad \dfrac{\sqrt{3}}{2} \cos q° + \dfrac{1}{2} \sin q°$

C $\quad \dfrac{2}{\sqrt{3}} \cos q° + \dfrac{1}{\sqrt{3}} \sin q°$

D $\quad \dfrac{2}{\sqrt{5}} \cos q° + \dfrac{1}{\sqrt{5}} \sin q°$

6. The vectors $\boldsymbol{p} = \begin{pmatrix} a \\ -1 \\ 2 \end{pmatrix}$ and $\boldsymbol{q} = \begin{pmatrix} -1 \\ a \\ 3 \end{pmatrix}$ are perpendicular. What is the value of a?

A $\quad -\dfrac{3}{2}$

B $\quad 0$

C $\quad \dfrac{5}{2}$

D $\quad 3$

7. The diagram shows a straight line graph with equation $y = f(x)$.

The line passes through the point $(0, 3)$.

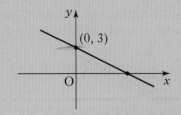

Which of the following diagrams could be the graph with equation $y = 3 - f(x)$?

A

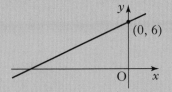

B

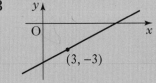

C

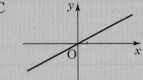

D

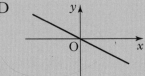

8. A sequence is defined by the recurrence relation $u_{n+1} = 0 \cdot 9 u_n + 90$
What is the limit of this sequence?

A -900

B $94 \cdot 5$

C 100

D 900

9. The diagram shows a circle, centre C(0,–3) with a tangent drawn at the point P(–2, 0).

   What is the equation of this tangent?

   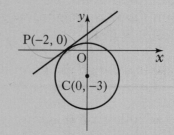

   A $y = \dfrac{2}{3}(x + 2)$

   B $y + 2 = -\dfrac{2}{3}x$

   C $y + 3 = -\dfrac{3}{2}x$

   D $y = \dfrac{3}{2}(x + 2)$

10. The equation $\sqrt{2}\cos\theta + 1 = 0$ has solution $\theta = \alpha$ where $\pi \le \alpha \le \frac{3\pi}{2}$. What is the value of α?

    A $\dfrac{3\pi}{4}$

    B $\dfrac{5\pi}{4}$

    C $\dfrac{4\pi}{3}$

    D $\dfrac{3\pi}{2}$

11. Find $\int 6\cos 2x\, dx$

    A $-12\sin 2x + c$

    B $3\sin 2x + c$

    C $-6\sin 2x + c$

    D $6\sin(x^2) + c$

12. If $f(x) = \sqrt{x^2 + 1}$ what is $f'(x)$?

A   1

B $\dfrac{x}{\sqrt{x^2 + 1}}$

C $\dfrac{x}{\left(\sqrt{x^2 + 1}\right)^3}$

D $2x\sqrt{x^2 + 1}$

13. In the diagram ABCD represents a tetrahedron.

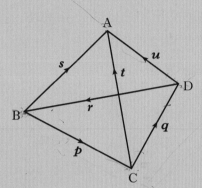

$\overrightarrow{BC}$ represents $\boldsymbol{p}$, $\overrightarrow{CD}$ represents $\boldsymbol{q}$,

$\overrightarrow{DB}$ represents $\boldsymbol{r}$, $\overrightarrow{BA}$ represents $\boldsymbol{s}$,

$\overrightarrow{CA}$ represents $\boldsymbol{t}$ and $\overrightarrow{DA}$ represents $\boldsymbol{u}$.

One of these statements is false, which one?

A $\boldsymbol{p} = -\boldsymbol{q} + \boldsymbol{s} - \boldsymbol{u}$

B $\boldsymbol{q} = -\boldsymbol{p} + \boldsymbol{s} + \boldsymbol{u}$

C $\boldsymbol{r} = -\boldsymbol{p} - \boldsymbol{t} + \boldsymbol{u}$

D $\boldsymbol{s} = \boldsymbol{p} + \boldsymbol{q} + \boldsymbol{u}$

14. P divides AB in the ratio 3:2 where A is the point $(-3, 2, 6)$ and B is the point $(7, -3, 1)$. What is the y-coordinate of P?

A -1

B   0

C   1

D   3

15. The diagram shows a graph with equation of the form $y = k(x - a)(x - b)$

What is the equation of the graph?

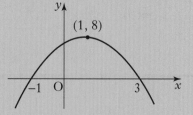

A $y = -2(x + 1)(x - 3)$

B $y = -2(x - 1)(x + 3)$

C $y = 8(x + 1)(x - 3)$

D $y = 8(x - 1)(x + 3)$

16. The graph shown in the diagram has equation of the form $y = k \times 2^{-x}$

What is the value of k?

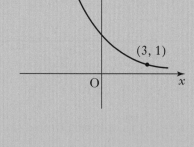

A $-\dfrac{3}{2}$

B $\dfrac{1}{8}$

C $\dfrac{1}{3}$

D 8

17. $3x^2 - 6x + 11$ is expressed in the form $3(x + a)^2 + b$
What is the value of b?

A   1

B   6

C   8

D   11

18. ABC is an equilateral triangle with side length 3 units. $\overrightarrow{AB}$ represents $\boldsymbol{v}$ and $\overrightarrow{AC}$ represents $\boldsymbol{w}$.

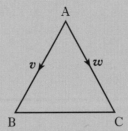

Find the value of $\boldsymbol{v} \cdot (\boldsymbol{v} - \boldsymbol{w})$

A   0

B $\dfrac{9}{2}$

C $\dfrac{9\sqrt{3}}{2}$

D   9

19. A function f is defined by $f(x) = \dfrac{5}{2(x^2 - 3x + 2)}$. A suitable domain for f is the set of Real numbers apart from which values?

A $x = -2$ and $x = -1$

B $x = 0$

C $x = 1$ and $x = 2$

D $x = 2$ and $x = 4$

20. The graph $y = 2\log_5(x + 3)$ is shown in the diagram.

At what point does this graph intersect the x-axis?

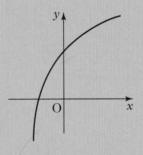

A $(-4, 0)$

B $(-3, 0)$

C $\left(-\dfrac{3}{2}, 0\right)$

D $(-2, 0)$

[End of section A]

SECTION B

Marks

21. The diagram shows a sketch of the curve with equation $y = \dfrac{1}{16}x^4 - \dfrac{1}{8}x^2 + x$. The line $y = x + c$ is a tangent to this curve.

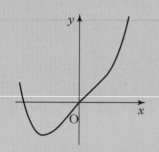

Find the possible values for c and for each value find the coordinates of the point of contact of the tangent.

7

22. A function f is defined by the formula $f(x) = x^3 + 3x^2 - 4$.

   (a) Find the coordinates of the stationary points on the graph with equation $y = f(x)$ and determine their nature.

6

   (b) (i) Show that $(x + 2)$ is a factor of $x^3 + 3x^2 - 4$.

      (ii) Hence or otherwise factorise $x^3 + 3x^2 - 4$ fully.

5

   (c) Find the coordinates of the points where the curve $y = f(x)$ crosses the x and y-axes and hence sketch the curve.

4

23. Functions f and g are defined by $f(x) = 2x - 1$ and $g(x) = \log_{12} x$ with suitable domains.

   (a) Show that the equation $f(g(x)) + g(f(x)) = 0$ has a solution $x = 2$.

6

   (b) Show that the equation has no other real solutions.

2

[End of section B]

[End of question paper]

Mathematics   Higher

Practice Papers
for SQA Exams

**Exam B
Higher
Paper 2**

You are allowed 1 hour, 10 minutes to complete this paper.

You may use a calculator.

Full marks will only be awarded where your answer includes relevant working.

You will not receive any marks for answers derived from scale drawings.

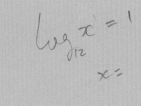

FORMULAE LIST

Trigonometric formulae

$$\sin (A \pm B) = \sin A \cos B \pm \cos A \sin B$$
$$\cos (A \pm B) = \cos A \cos B \mp \sin A \sin B$$
$$\sin 2A = 2\sin A \cos A$$
$$\cos 2A = \cos^2 A - \sin^2 A$$
$$= 2\cos^2 A - 1$$
$$= 1 - 2\sin^2 A$$

Circle

The equation $x^2 + y^2 + 2nx + 2py + c = 0$ represents a circle centre $(-n, -p)$ and radius $\sqrt{n^2 + p^2 - c}$.

The equation $(x - a)^2 + (y - b)^2 = r^2$ represents a circle centre (a, b) and radius r.

Table of standard integrals

$f(x)$	$\int f(x)dx$
$\sin ax$	$-\dfrac{1}{a}\cos ax + C$
$\cos ax$	$\dfrac{1}{a}\sin ax + C$

Table of standard derivatives

$f(x)$	$f'(x)$
$\sin ax$	$a \cos ax$
$\cos ax$	$-a \sin ax$

Scalar Product　　$a.b = |a||b| \cos \theta$, where θ is the angle between a and b

or　　$a.b = a_1b_1 + a_2b_2 + a_3b_3$ where $a = \begin{pmatrix} a_1 \\ a_2 \\ a_3 \end{pmatrix}$ and $b = \begin{pmatrix} b_1 \\ b_2 \\ b_3 \end{pmatrix}$.

1. Three circles have equations as follows:

$$\text{Circle A} : x^2 + y^2 + 4x - 6y + 5 = 0$$

$$\text{Circle B} : (x - 2)^2 + (y + 1)^2 = 2$$

$$\text{Circle C} : (x - 2)^2 + (y + 1)^2 = 40$$

(*a*) (i) State the centre of circle A. **1**

(ii) Show that the radius of circle A is $2\sqrt{2}$. **1**

(*b*) (i) Calculate the distance between the centres of circles A and B writing your answer as a surd in its simplest form. **2**

(ii) Hence show that circles A and B do not intersect. **2**

(*c*) Circles A and C intersect at points P and Q. Chord PQ has equation $y = x + 5$. Find the coordinates of points P and Q if P lies to the left of Q. **5**

2. The diagram shows kite ABCD with diagonal AC drawn. The vertices of the kite are A(−5, 2), B(−3, 8), C(3, 6) and D(5, −8).

The dotted line shows the perpendicular bisector of AB.

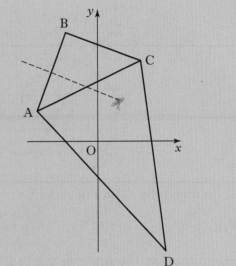

(*a*) Show that the perpendicular bisector of AB has equation $3y + x = 11$. **4**

(*b*) Find the equation of the median from C in triangle ACD. **3**

(*c*) The perpendicular bisector of AB and the median from C in triangle ACD meet at the point S. Find the coordinates of S. **3**

3. Solve the equation

$$3 \cos 2x° + 9 \cos x° = \cos^2 x° - 7 \text{ for } 0 \le x < 360$$

5

4. OABC, DEFG is a cuboid.

The vertex F is the point (5, 6, 2).

M is the midpoint of DG.

N divides AB in the ratio 1:2.

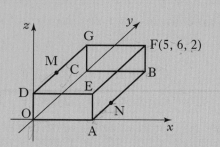

(a) Find the coordinates of M and N.

2

(b) Write down the components of $\overrightarrow{MB}$ and $\overrightarrow{MN}$.

2

(c) Find the size of angle BMN.

5

5. The diagram below shows the graphs $y = f(x)$ and $y = g(x)$ where $f(x) = m \sin x$ and $g(x) = n \cos x$

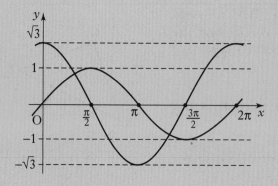

(a) Write down the values of m and n.

2

(b) Write $f(x) - g(x)$ in the form $k \sin(x - a)$ where $k > 0$ and $0 < a < \dfrac{\pi}{2}$.

4

(c) Hence find, in the interval $0 \le x \le \pi$ the x-coordinate of the point on the curve $y = f(x) - g(x)$ where the gradient is 2.

2

6. The diagram shows the graph with equation $y = x^4 - 1$.

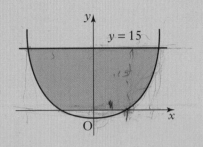

The graph has the y-axis as an axis of symmetry.

The shaded area lies between the curve, the x-axis and the line $y = 15$.

Calculate the exact value of the shaded area.

Marks

8

7. The diagram shows a parabola with equation $y = f(x)$ passing through the points $(-3, 0)$, $(0, 9)$ and $(3, 0)$.

OAPB is a rectangle with A and B lying on the axes and P lying on the parabola as shown. $OA = m$, $0 \le m \le 3$.

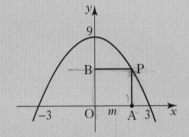

(*a*) If $f(x)$ is of the form $-x^2 + a$ where a is a constant, determine the value of a.   1

(*b*) Show that $AP = 9 - m^2$.   1

(*c*) Find the value of m for which the area of the rectangle has a maximum.   6

(*d*) Find the exact value of this maximum area.   1

[End of question paper]

Practice Exam C

Mathematics　　　　Higher

Practice Papers
for SQA Exams

Exam C
Higher
Paper 1
Non-calculator

You are allowed 1 hour, 30 minutes to complete this paper.

You must not use a calculator.

Full marks will only be awarded where your answer includes relevant working.

You will not receive any marks for answers derived from scale drawings.

FORMULAE LIST

Trigonometric formulae

$$\sin (A \pm B) = \sin A \cos B \pm \cos A \sin B$$
$$\cos (A \pm B) = \cos A \cos B \mp \sin A \sin B$$
$$\sin 2A = 2\sin A \cos A$$
$$\cos 2A = \cos^2 A - \sin^2 A$$
$$= 2\cos^2 A - 1$$
$$= 1 - 2\sin^2 A$$

Circle

The equation $x^2 + y^2 + 2nx + 2py + c = 0$ represents a circle centre $(-n, -p)$ and radius $\sqrt{n^2 + p^2 - c}$.

The equation $(x - a)^2 + (y - b)^2 = r^2$ represents a circle centre (a, b) and radius r.

Table of standard integrals

$f(x)$	$\int f(x)dx$
$\sin ax$	$-\dfrac{1}{a}\cos ax + C$
$\cos ax$	$\dfrac{1}{a}\sin ax + C$

Table of standard derivatives

$f(x)$	$f'(x)$
$\sin ax$	$a \cos ax$
$\cos ax$	$-a \sin ax$

Scalar Product $a.b = |a||b| \cos \theta$, where θ is the angle between a and b

or $a.b = a_1b_1 + a_2b_2 + a_3b_3$ where $a = \begin{pmatrix} a_1 \\ a_2 \\ a_3 \end{pmatrix}$ and $b = \begin{pmatrix} b_1 \\ b_2 \\ b_3 \end{pmatrix}$.

1. What is the limit of the sequence generated by the recurrence relation $u_{n+1} = 0 \cdot 9\, u_n - 1$?

   A -10

   B $-\dfrac{10}{9}$

   C $\dfrac{20}{9}$

   D 10

2. The line through points A$(2k, 3)$ and B$(k, 5)$ has gradient 4. What is the value of k?

   A -2

   B $-\dfrac{1}{2}$

   C $\dfrac{2}{3}$

   D 8

3. The following two statements are true for the polynomial $f(x)$:

   (1) $x^2 - 4$ is a factor of $f(x)$

   (2) $f(-1) = 0$

   Which of the following is a possible expression for $f(x)$?

   A $(x - 1)^2(x - 2)$

   B $(x - 2)(x + 2)$

   C $(x - 1)(x - 2)^2$

   D $(x + 1)(x - 2)(x + 2)$

4. A sequence is defined by the recurrence relation $u_{n+1} = -\dfrac{1}{2}u_n + 1$, $u_0 = 4$; what is the value of u_2?

   A $-\dfrac{1}{2}$

   B $\dfrac{3}{4}$

   C $\dfrac{3}{2}$

   D $\dfrac{5}{2}$

5. Solve the equation $2 \cos x - \sqrt{2} = 0$ for $0 \leq x \leq \frac{\pi}{2}$

A $\dfrac{\pi}{6}$

B $\dfrac{\pi}{4}$

C $\dfrac{3\pi}{4}$

D $\dfrac{5\pi}{4}$

6. What is the gradient of the line perpendicular to the line with equation $4y = -3x + 2$?

A -4

B -2

C $\dfrac{4}{3}$

D   7

7. The points E(1, −1, −1), F(−1, −1, 0) and G(−7, −1, 3) are collinear. In what ratio does F divide EG?

A   2:1

B   −1:3

C   1:3

D   3:1

8. What is the remainder when $2x^4 - 3x^3 - 3x + 1$ is divided by $x - 2$?

A   −21

B   −9

C   −3

D   3

9. $(x + 4)(x - 2)$ can be written in the form $(x + a)^2 + b$. What is the value of b?

A   −12

B   −9

C   −8

D   1

10. A function is defined by $f(x) = (1 - x^3)^{\frac{1}{3}}$. Find $f'(x)$.

A $-x^2(1 - x^3)^{-\frac{2}{3}}$

B $(1 - x^3)^{-\frac{2}{3}}$

C $(1 - 3x^2)^{\frac{1}{3}}$

D $-4(1 - x^3)^{\frac{4}{3}}$

11. The diagram shows the graph with equation $y = f(x)$. Which of the following shows the graph with equation $y = -(f(x) + 1)$?

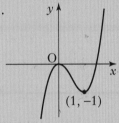

A

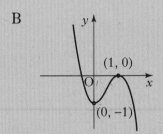

B

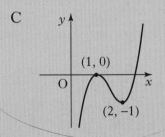

C

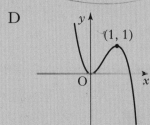

D

12. M is the midpoint of the line AB where A and B have coordinates $(-1, 2, 0)$ and $(-2, 3, 1)$ respectively. What is the position vector of M?

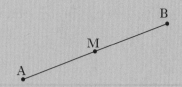

A $\begin{pmatrix} 3/2 \\ -5/2 \\ -1/2 \end{pmatrix}$

B $\begin{pmatrix} -3/2 \\ 5/2 \\ 1/2 \end{pmatrix}$

C $\begin{pmatrix} 1/2 \\ -1/2 \\ -1/2 \end{pmatrix}$

D $\begin{pmatrix} 1 \\ -1 \\ -1 \end{pmatrix}$

13. $f'(x) = x^2 + 1$ for a function f. Which statement is true?

A f has no stationary points.

B f has exactly one stationary point.

C f has exactly two stationary points.

D f has more than two stationary points.

14. The diagram shows the line OA where A is the point with coordinates $(2,3)$. $a°$ is the angle between OA and the positive direction of the x-axis. Which of the following gives the value of a?

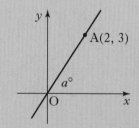

A $-\tan^{-1}\dfrac{2}{3}$

B $\tan^{-1}\dfrac{2}{3}$

C $-\tan^{-1}\dfrac{3}{2}$

D $\tan^{-1}\dfrac{3}{2}$

15. Which of the following expressions gives a simplification of
$\log_4 (x^2 - 4) - 2 \log_4 (x - 2)$?

A   0

B $\log_4 x(x - 2)$

C $\log_4 \dfrac{x + 2}{x - 2}$

D $\log_4 (x - 2)^2(x + 2)$

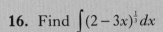

16. Find $\int (2 - 3x)^{\frac{1}{3}} dx$

A $-\dfrac{1}{4}(2 - 3x)^{\frac{4}{3}} + c$

B $\dfrac{1}{3}(2 - 3x)^{-\frac{2}{3}} + c$

C $9(2 - 3x)^{-\frac{1}{3}} + c$

D $(2 - x) + c$

17. Points P and Q have coordinates $(-1, 2, 5)$ and $(-3, -1, 4)$ respectively. If $\overrightarrow{QR} = -2\overrightarrow{PQ}$ what are the coordinates of R?

A $(1, 5, 6)$

B $(5, -3, -14)$

C $(-7, -7, 2)$

D $(-8, -2, -10)$

18. A function f is defined by $f(x) = -2 \sin 3x$. Find $f'(x)$.

A $6 \sin 3x$

B $-\dfrac{2}{3} \cos 3x$

C $-6 \cos 3x$

D $-2 \sin 3$

19. The graph shows a parabola with equation of the form $y = k(x - 1)(x + 2)$. What is the value of k?

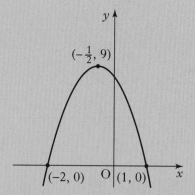

A $-\dfrac{36}{5}$

B -4

C $-\dfrac{1}{176}$

D 9

20. Here is an integration formula:

$$\int \frac{1}{(4x+1)^{3/2}}\,dx = -\frac{1}{2(4x+1)^{1/2}} + c$$

Use this formula to calculate the exact value of $\displaystyle\int_0^2 \frac{1}{(4x+1)^{3/2}}\,dx$.

A -8

B $-\dfrac{26}{27}$

C $\dfrac{1}{6}$

D $\dfrac{1}{3}$

[End of section A]

SECTION B

Marks

21. Solve the equation $\sin 2x - \sqrt{3}\sin x = 0$ for $0 \le x \le 2\pi$

5

22. (a) Find the stationary points on the curve with equation $y = x^3 - 3x^2 - 24x - 28$ and justify their nature.

7

   (b) The curve intersects the x-axis at $(7,0)$. Sketch the curve.

2

23. ABCD is a rectangle with point P lying on side AB, 1 unit from A. AD = 2 units.

The dotted line PD on the diagram shows the bisector of angle APC with angle APD = angle DPC = $y°$.

   (a) Find $\theta°$ in terms of $y°$.

1

   (b) Hence find the exact value of $\sin \theta°$.

6

24. Solve the equation
$\log_{\sqrt{2}} x - \log_{\sqrt{2}} 2 = 2$

4

25. The diagram shows the cubic graph with equation $y = x^2(x - 1)$. A function f is such that $y = f'(x)$ is the same as the graph shown in the diagram.

If $f(2) = \dfrac{1}{3}$, find the formula for $f(x)$.

5

[End of section B]

[End of question paper]

Mathematics   Higher

Practice Papers
for SQA Exams

**Exam C
Higher
Paper 2**

You are allowed 1 hour, 10 minutes to complete this paper.

You may use a calculator.

Full marks will only be awarded where your answer includes relevant working.

You will not receive any marks for answers derived from scale drawings.

Scotland's leading educational publishers

FORMULAE LIST

Trigonometric formulae

$$\sin(A \pm B) = \sin A \cos B \pm \cos A \sin B$$
$$\cos(A \pm B) = \cos A \cos B \mp \sin A \sin B$$
$$\sin 2A = 2\sin A \cos A$$
$$\cos 2A = \cos^2 A - \sin^2 A$$
$$= 2\cos^2 A - 1$$
$$= 1 - 2\sin^2 A$$

Circle

The equation $x^2 + y^2 + 2nx + 2py + c = 0$ represents a circle centre $(-n, -p)$ and radius $\sqrt{n^2 + p^2 - c}$.

The equation $(x - a)^2 + (y - b)^2 = r^2$ represents a circle centre (a, b) and radius r.

Table of standard integrals

$f(x)$	$\int f(x)dx$
$\sin ax$	$-\dfrac{1}{a}\cos ax + C$
$\cos ax$	$\dfrac{1}{a}\sin ax + C$

Table of standard derivatives

$f(x)$	$f'(x)$
$\sin ax$	$a \cos ax$
$\cos ax$	$-a \sin ax$

Scalar Product $a.b = |a||b|\cos\theta$, where θ is the angle between a and b

or $a.b = a_1 b_1 + a_2 b_2 + a_3 b_3$ where $a = \begin{pmatrix} a_1 \\ a_2 \\ a_3 \end{pmatrix}$ and $b = \begin{pmatrix} b_1 \\ b_2 \\ b_3 \end{pmatrix}$.

Marks

1. (a) Express $3\sin x° - \cos x°$ in the form $k\sin(x - a)°$ where $k > 0$ and $0 \le a \le 90$. 4

   (b) Hence solve the equation $3\sin x° - \cos x° = 1$ for $0 \le x \le 90$. 3

2. The vectors $\overrightarrow{BA}$ and $\overrightarrow{BC}$ have components $\begin{pmatrix} -2 \\ 3 \\ 5 \end{pmatrix}$ and $\begin{pmatrix} 1 \\ -1 \\ 3 \end{pmatrix}$ respectively. Calculate the size of angle ABC. 5

3. Prove that for all values of c the equation $x^2 - 2x + c^2 + 2 = 0$ has no real roots. 4

4. The diagram shows the graph with equation $y = \frac{1}{3}x^3 - 2x^2$.

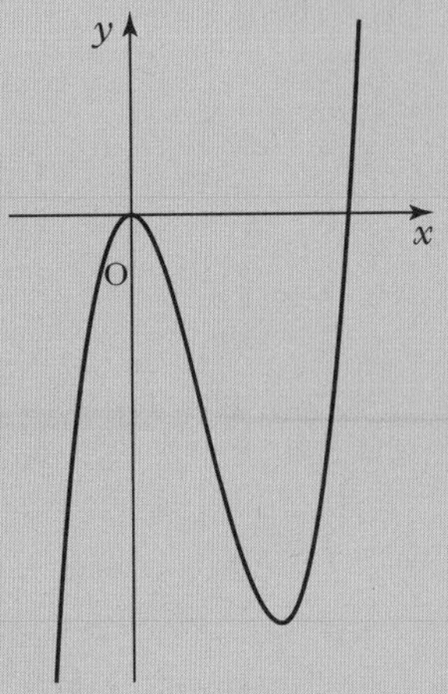

   (a) A tangent to this curve has gradient -4. Find the x-coordinate of the point of contact. 5

   (b) Hence find the equation of this tangent. 2

5. A drug is given to a patient. The concentration, C_t milligrams per millilitre (mg/ml), of the drug in the patient's blood t hours after it is administered is given by the formula:

$$C_t = C_0 e^{-\frac{t}{4}}$$

   where C_0 is the concentration in the blood immediately after the drug was administered.

   (a) If the concentration is $3·5$ mg/ml after 3 hours, what was the concentration of the drug just after it was administered? 3

   (b) In general, after a dose of this drug has been administered, how long does it take for the initial concentration to halve? 4

6. The line L is a tangent to the circle with centre C_1 and equation

$$x^2 + y^2 - 4x - 6y + 8 = 0.$$

The point of contact A has coordinates $(1,5)$.

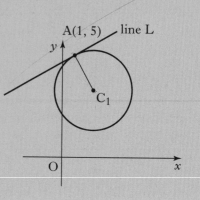

(a) Show that the equation of line L is $2y - x = 9$.

Marks

The circle with centre C_2 has equation

$$x^2 + y^2 + 2x + 2y - 18 = 0.$$

4

(b) Show that line L is also a tangent to this circle.

(c) If B is the point of contact, find the exact length of AB.

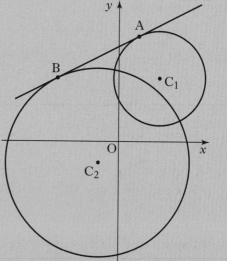

5

2

7. The graph shows a cubic function with equation $y = f(x)$.

The graph has stationary points at $(-2, 72)$ and $\left(\frac{8}{3}, -\frac{800}{27}\right)$.

The graph intersects the axes at the points $(-4, 0)$, $(1, 0)$, $(4, 0)$ and $(0, 32)$.

Sketch the graph of $y = f'(x)$.

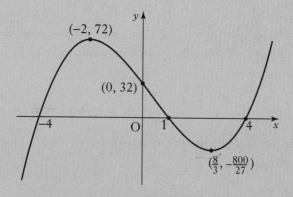

3

8. An open box is in the shape of a cuboid and was made from a sheet of tin.

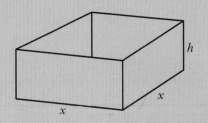

The box has a square base of side x cm and a height of h cm. The volume of the box is $62\frac{1}{2}$cm^3

(*a*) Show that the Area, A cm^2, of tin required to make the box is given by
$$A(x) = \frac{250}{x} + x^2.$$

3

(*b*) Find the value of x for which this area is a minimum.

5

9. The diagram shows a rectangular metal plate with dimensions 5cm × 6cm.

The plate has a parabolic section removed from it.

The equation of the parabola used to make this section is $y = x^2 - 6x + 10$.

The scale of the diagram is 1 unit = 1 cm.

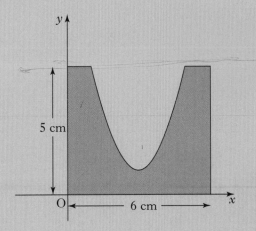

Find the shaded area, in square centimetres, of the metal plate.

8

Marks

[End of question paper]

Worked Answers

1.
$$\sin 2y° = 2 \sin y° \cos y°$$

$$= 2 \times \frac{2}{2\sqrt{3}} \times \frac{2\sqrt{2}}{2\sqrt{3}}$$

$$= \frac{2 \times 2 \times 2 \times \sqrt{2}}{2 \times \sqrt{3} \times 2 \times \sqrt{3}}$$

$$= \frac{2\sqrt{2}}{3}$$

Choice C

2 marks

- Use of the Double angle formula. The formula is given to you on the formula sheet:

$$\sin 2A = 2 \sin A \cos A$$

- Use of SOHCAHTOA in the right angled triangle.

- Remember $\sqrt{a} \times \sqrt{a} = a$. In this case $\sqrt{3} \times \sqrt{3} = 3$.

- Multiplying fractions: $\frac{a}{b} \times \frac{c}{d} = \frac{a \times c}{b \times d}$;

HMRN: p 35–36

2.
$$y = \frac{x^4 + 1}{x} = \frac{x^4}{x} + \frac{1}{x}$$

$$= x^3 + x^{-1}$$

$$\frac{dy}{dx} = 3x^2 - x^{-2}$$

$$= 3x^2 - \frac{1}{x^2}$$

Choice D

2 marks

- Before differentiating you must 'split the fraction' to get powers of x: x^3 and x^{-1}.

- Differentiation rule:

$$y = ax^n \Rightarrow \frac{dy}{dx} = nax^{n-1}$$

- Remember $a^{-n} = \frac{1}{a^n}$

HMRN: p 18–19

3.
Least value of $2(x + 2)^2 - 1$
is $2 \times 0^2 - 1$ when $x + 2 = 0$
i.e. Least value is -1 when $x = -2$
Stationary point is $(-2,-1)$

Choice C

2 marks

- The general results are:

$a > 0$ $a(x + b)^2 + c$

minimum value is c when $x = -b$

$a < 0$ $a(x + b)^2 + c$

maximum value is c when $x = -b$

- Remember a number squared has to be positive or zero – never negative.

HMRN: p 13–14

4. $f(g(x)) = f(3 - x) = \sqrt{3 - x}$

Choice B

2 marks

- Always work from the inside out so $g(x)$ is replaced first by $3-x$ then f is used – the square root.

HMRN: p 10

5.

$$\cos(p+q)^\circ = \cos p^\circ \cos q^\circ \\ - \sin p^\circ \sin q^\circ$$

$$= \frac{1}{\sqrt{5}} \times \frac{3}{\sqrt{13}} - \frac{2}{\sqrt{5}} \times \frac{2}{\sqrt{13}}$$

$$= \frac{3}{\sqrt{65}} - \frac{4}{\sqrt{65}} = \frac{3-4}{\sqrt{65}}$$

$$= -\frac{1}{\sqrt{65}}$$

Choice B

2 marks

- Notice there are two right-angled triangles:

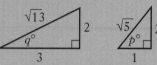

- The addition formula is given on your formula sheet:

$$\cos(A \pm B) = \cos A \cos B \mp \sin A \sin B$$

HMRN: p 35

6.

$$\int \frac{3}{\sqrt{x}} dx = \int \frac{3}{x^{\frac{1}{2}}} dx = \int 3x^{-\frac{1}{2}} dx$$

$$= \frac{3x^{\frac{1}{2}}}{\frac{1}{2}} + c = 6x^{\frac{1}{2}} + c$$

$$= 6\sqrt{x} + c$$

Choice A

2 marks

- The formula being used here is:

$$\int ax^n = \frac{ax^{n+1}}{n+1} + c$$

- It is important to get the $\sqrt{x}$ or $x^{\frac{1}{2}}$ up from the bottom of the fraction. To do this you use $\frac{1}{x^n} = x^{-n}$

- Notice: $\dfrac{3x^{\frac{1}{2}} \ (\times 2)}{\frac{1}{2} \ (\times 2)} = \dfrac{6x^{\frac{1}{2}}}{1} = 6x^{\frac{1}{2}}$

HMRN: p 31

7.

$$x^2 + y^2 - 2x + 6y - 1 = 0$$

Centre: $(1, -3)$

Radius: $\sqrt{1^2 + (-3)^2 - (-1)}$

$$= \sqrt{1 + 9 + 1} = \sqrt{11}$$

Choice B

2 marks

- Here is the process you are using:

$$x^2 + y^2 + \overset{\frown}{2g}x + \overset{\frown}{2f}y + c = 0$$

halve and change sign   subtract

Centre: $(-g, \quad -f)$

square and add

Radius: $\sqrt{(-g)^2 + (-f)^2 - c}$

- In this case $c = -1$ – not 1
- $(-3)^2 = 9$. Squaring always gives you a positive answer or zero.

HMRN: p 39

8.

$$P(-1, -3, 5) \text{ and } Q(0, 5, -2)$$

PQ

$$= \sqrt{(-1-0)^2 + (-3-5)^2 + (5-(-2))^2}$$

$$= \sqrt{(-1)^2 + (-8)^2 + 7^2}$$

$$= \sqrt{1 + 64 + 49} = \sqrt{114}$$

Choice D

2 marks

- You use the distance formula:
  $A(x_1, y_1, z_1,)$ and $B(x_2, y_2, z_2)$

$$AB = \sqrt{(x_1 - x_2)^2 + (y_1 - y_2)^2 + (z_1 - z_2)^2}$$

x difference y difference z difference

- Watch that all your squarings: $(-1)^2$ and $(-8)^2$ give positive answers: 1 and 64.
- No calculator is allowed in this paper so be careful with the calculations: $5-(-2) = 5 + 2$ etc.

HMRN: p 7

9.

$u_{n+1} = u_n^2 - 1$, $u_0 = -2$

So $u_1 = (-2)^2 - 1 = 4 - 1 = 3$

and $u_2 = 3^2 - 1 = 9 - 1 = 8$

Choice C

2 marks

- Here is the process:

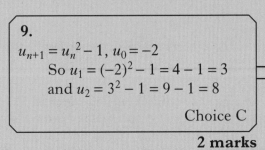

square and   square and
subtract 1   subtract 1

so

$-2 \quad\quad 3 \quad\quad 8$

$(-2)^2 - 1 \quad 3^2 - 1$

HMRN: p 23

10.

$y = f(x) \rightarrow y = f(x - 4)$

This is a translation of 4 units in the positive direction of the x-axis

$y = f(x - 4) \rightarrow y = -f(x - 4)$

followed by reflection in the y-axis. Check the turning point:

$(-2, 5) \rightarrow (2, 5) \rightarrow (2, -5)$

Choice A

2 marks

- It is useful to remember similar results: $y = \sin(x - 30)°$ the $y = \sin x°$ graph moves 30° to the right (parallel to the x-axis).

$y = -\sin x°$ The $y = \sin x°$ graph is 'flipped' in the x-axis.

- From the graph $f(-2) = 5$ now consider $y = -f(x - 4)$. Choice A has $x = 2$ so $y = -f(2 - 4) = -f(-2) = -5$ as required so this further check confirms choice A.

HMRN: p 9–10

11.

$$3x^2 + 1x + m = 0$$

Compare $ax^2 + bx + c = 0$

Discriminant $= b^2 - 4ac$

$\quad\quad\quad = 1^2 - 4 \times 3 \times m$

$\quad\quad\quad = 1 - 12m$

For equal roots: Discriminant $= 0$

$\Rightarrow 1 - 12m = 0 \Rightarrow 1 = 12m \Rightarrow m = \dfrac{1}{12}$

Choice C

2 marks

- The general result here is:

$$ax^2 + bx + c = 0$$

if

$b^2 - 4ac > 0$: 2 Real roots (distinct)

$b^2 - 4ac = 0$: 1 Real root (equal)

$b^2 - 4ac < 0$: 0 Real roots

- $1 = 12m$ divide both sides by 12

HMRN: p 27–28

12.

$x^2 + y^2 + 2x - 4y - 5 = 0$

Centre: $C(-1, 2)$

$$m_{AC} = \frac{3-2}{2-(-1)} = \frac{1}{3}$$

$\Rightarrow m_\perp = -3$

gradient of
tangent is -3

Choice B

2 marks

- You should always draw a sketch in a question like this. The sketch will usually suggest the strategy to use.

- The strategy here is to use the fact that the tangent is perpendicular to the radius CA.

- Remember, if two lines have gradients m_1 and m_2 and are perpendicular then

  $m_1 \times m_2 = -1$ This gives $m_2 = -\frac{1}{m_1}$.

  The process is "invert and change the sign". In this case $\frac{1}{3}$ changes to $-\frac{3}{1}$.

HMRN: p 4

13.

$7 - 8x - x^2$

$\quad = -[x^2 + 8x - 7]$

$\quad = -[(x+4)(x+4) - 16 - 7]$

$\quad = -[(x+4)^2 - 23]$

$\quad = -(x+4)^2 + 23$

$\quad = 23 - (x+4)^2$

Compare $a - (x+b)^2$

So $a = 23$

Choice D

2 marks

- When the x^2 term is negative you take out -1 as a common factor. In this case this allows you to deal with $x^2 + 8x - 7$ which is an easier problem.

- Once you have $x^2 + 8x - 7 = (x+4)^2 - 23$ you can then multiply each term by -1, the factor you removed at the start.

- An alternative method is:

  $7 - [x^2 + 8x]$

  $= 7 - [(x+4)(x+4) - 16]$

  $= 7 + 16 - (x+4)^2$

  $= 23 - (x+4)^2$

HMRN: p 13

14.

$$\overrightarrow{AB} = \boldsymbol{b} - \boldsymbol{a} = \begin{pmatrix} 6 \\ 1 \\ 1 \end{pmatrix} - \begin{pmatrix} 10 \\ -1 \\ 3 \end{pmatrix} = \begin{pmatrix} -4 \\ 2 \\ -2 \end{pmatrix}$$

$$\overrightarrow{BC} = \boldsymbol{c} - \boldsymbol{b} = \begin{pmatrix} 4 \\ 2 \\ t \end{pmatrix} - \begin{pmatrix} 6 \\ 1 \\ 1 \end{pmatrix} = \begin{pmatrix} -2 \\ 1 \\ t-1 \end{pmatrix}$$

So $\overrightarrow{BC} = \frac{1}{2}\overrightarrow{AB}$

relationship holds for
z-components:

$$t - 1 = \frac{1}{2} \times (-2) \Rightarrow t - 1 = -1 \Rightarrow t = 0$$

Choice A

2 marks

- If A, B and C are collinear then $\overrightarrow{AB}$ and $\overrightarrow{BC}$ are connected in the following manner: $\overrightarrow{BC} = k\overrightarrow{AB}$.

  In this case $k = \frac{1}{2}$. This can be determined using the x-coordinates: $-2 = k \times (-4)$ and checked using the y-coordinates: $1 = k \times 2$.

- Alternatively you could use $\overrightarrow{AB} = 2 \times \overrightarrow{BC}$ giving $-2 = 2(t-1) \Rightarrow -2 = 2t - 2 \Rightarrow 2t = 0 \Rightarrow t = 0$.

HMRN: p 44–45

15.

$$\frac{dy}{dx} = 4x^3 - x^2 - 1$$

$$\Rightarrow y = \int (4x^3 - x^2 - 1)\, dx$$

$$\Rightarrow y = \frac{4x^4}{4} - \frac{x^3}{3} - x + c$$

$$\Rightarrow y = x^4 - \frac{1}{3}x^3 - x + c$$

$y = 0$ when $x = 0$ since $(0,0)$ lies on the curve.

$$\Rightarrow 0 = 0^4 - \frac{1}{3} \times 0^3 - 0 + c \Rightarrow c = 0$$

So $y = x^4 - \frac{1}{3}x^3 - x$

Choice A

2 marks

- Going from y to $\frac{dy}{dx}$ is 'differentiation'. The inverse process, going from $\frac{dy}{dx}$ to y is 'integration'.

- The 'constant of integration' c is a crucial part of this question. The fact that the curve passes through the origin allows you to calculate c ($c = 0$).

- Notice $\int 4x^3 dx = \frac{{}^1 4x^4}{1\,4} + c = x^4 + c$ with cancellation of the two factors 4.

HMRN: p 31

16.

$$\log_2 9 = 3 - \log_2 x$$

$$\Rightarrow \log_2 9 + \log_2 x = 3$$

$$\Rightarrow \log_2 9x = 3$$

$$\Rightarrow 2^3 = 9x \Rightarrow 8 = 9x$$

$$\Rightarrow x = \frac{8}{9}$$

Choice B

2 marks

- The rules used here are:

$$\log_a m + \log_a n = \log_a mn$$

and $\quad \log_b a = c \quad \Leftrightarrow \quad b^c = a$
$\qquad$ (logarithmic form) $\qquad\qquad$ (exponential form)

HMRN: p 50

17.

$$\int_0^{\pi/3} \cos \frac{1}{2} x \, dx$$

$$= \left[\frac{\sin \frac{1}{2}x}{\frac{1}{2}} \right]_0^{\pi/3} = \left[2\sin\frac{1}{2}x \right]_0^{\pi/3}$$

$$= 2\sin\frac{1}{2} \times \frac{\pi}{3} - 2\sin\frac{1}{2} \times 0$$

$$= 2\sin\frac{\pi}{6} - 2\sin 0 = 2 \times \frac{1}{2} - 0 = 1$$

Choice C

2 marks

- The rules used here are:

$$\int \cos ax = \frac{\sin ax}{a} + c$$

and $[F(x)]_b^c = F(c) - F(b)$

- Multiplication of fractions:

$\frac{1}{2} \times \frac{\pi}{3} = \frac{1 \times \pi}{2 \times 3}$

- Exact value of $\sin \frac{\pi}{6}$ is obtained by using SOHCAHTOA in the 'half an equilateral triangle' diagram.

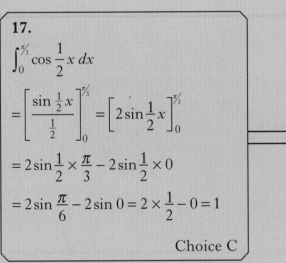

HMRN: p 49

18.

$q.(p - r) = q.p - q.r$

$= |q||p|\cos 120° - |q||r|\cos 120°$

$= 0$ since $|q| = |p| = |r|$

$q.(p + r) = q.p + q.r$

$= |q||p|\cos 120° + |q||r|\cos 120$

$\neq 0$

So (1) is true and (2) is false

Choice B

2 marks

- The results used in this question are:

$a.(b \pm c) = a.b \pm a.c$

and

$v.w = |v||w|\cos\theta$

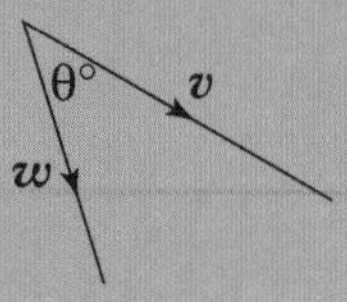

- The angle between q and p is 120° as is the angle between q and r. Remember for angles the vectors must come *out* from the vertex. The diagram shows an extension of the equilateral triangle to show the 120° angles.

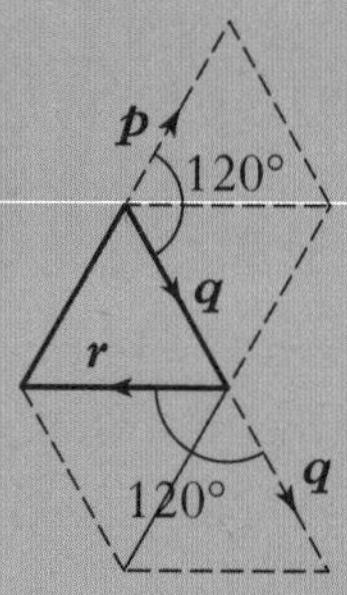

HMRN: p 47

19.

The period of the graph is $\dfrac{2\pi}{3}$.

Now $3 \times \dfrac{2\pi}{3} = 2\pi$ so there are 3 cycles compared to 1 cycle for $y = \cos x$.

The amplitude is $\dfrac{1}{2}$

The equation is $y = \dfrac{1}{2}\cos 3x$

(amplitude) (3 cycles)

Choice D

2 marks

- The results used here are:

$y = k\cos x$ k is the amplitude

and $y = k\cos nx$ period is $\frac{2\pi}{n}$ so there will be n cycles in the interval 0 to 2π.

- Choice A has the correct amplitude but would give $\frac{1}{3}$ of a cycle from 0 to 2π. Choice B has correct period $\frac{2\pi}{3}$ but the whole graph would be moved $\frac{1}{2}$ unit up parallel to the y-axis.

Choice C has the wrong amplitude $\left(\frac{2}{3}\right)$ and the wrong period ($\frac{1}{2}$ of a cycle from 0 to 2π).

HMRN: p16–17

20.

$\dfrac{\log_k 4}{\log_e 2} = 2e^0$

$\Rightarrow \dfrac{\log_k 4}{\log_e 2} = 2 \times 1 = 2$

$\Rightarrow \log_k 4 = 2\log_e 2 = \log_e 2^2$

So $\log_k 4 = \log_e 4$

$\Rightarrow k = e$

Choice D

2 marks

- The results used in this question are:

$a^0 = 1$ (in this case $e^0 = 1$)

$n\log_a b = \log_a b^n$
(in this case $2\log_e 2 = \log_e 2^2$).

- Adding or subtracting logs can be simplified, e.g. $\log_a m \pm \log_a n = \log_a mn$ or $\log_a \frac{m}{n}$
(addition)   (subtraction)

but $\dfrac{\log_a m}{\log_a n}$ cannot be simplified.

HMRN: p 50–51

21. (a)

$y = x^3 - 3x^2 + 4$ ✓

$\Rightarrow \dfrac{dy}{dx} = 3x^2 - 6x$ ✓

For stationary points set $\dfrac{dy}{dx} = 0$ ✓

$\Rightarrow 3x^2 - 6x = 0$

$\Rightarrow 3x(x - 2) = 0 \Rightarrow x = 0$ or 2 ✓

$\dfrac{dy}{dx} = 3x(x - 2):$

$x:$		0		2	
	+		−		+

Shape of graph: ╱ ╲ ╱ ✓

Nature: max min

when $x = 0$ $y = 0^3 - 3 \times 0^2 + 4 = 4$

So $(0, 4)$ is a maximum ✓

stationary point.

When $x = 2$ $y = 2^3 - 3 \times 2^2 + 4 = 0$

So $(2, 0)$ is a minimum ✓

stationary point.

7 marks

Differentiate
- 1 mark for knowing to differentiate.
- 1 mark for correctly differentiating.

Strategy
- Setting $\frac{dy}{dx} = 0$ allows you to home in on the x-coordinates of the stationary points.

Solving Equation
- The common factor here is $3x$.

Justification
- "justify their nature" means that you need a 'nature table' as shown in the solution.

y-coordinates
- 1 mark is allocated to the calculation of the two corresponding y-coordinates (4 and 0).

Interpretation
- Clear statements should be made about the type (max or min) of each stationary point.

HMRN: p 20–21

21. (b) (i)

$(x + 1)(x - 2)^2$

$= (x + 1)(x - 2)(x - 2)$

$= (x + 1)(x^2 - 4x + 4)$

$= x^3 - 4x^2 + 4x + x^2 - 4x + 4$ ✓

$= x^3 - 3x^2 + 4$

1 mark

Expand
- Since the answer $x^3 - 3x^2 + 4$ is given it is important that you show clearly all your steps in the expansion of the brackets.
- Notice for $(x + 1)(x^2 - 4x + 4)$ you have: $x(x^2 - 4x + 4)$ and $1(x^2 - 4x + 4)$.

21. (b) (ii)

For x-intercepts set

$y = 0$

So $(x + 1)(x - 2)^2 = 0$

$\Rightarrow x + 1 = 0$ or $x - 2 = 0$

$\Rightarrow x = -1$ or $x = 2$.

Intercepts are $(-1, 0), (2, 0)$

For y-intercepts set $x = 0$

So $y = 0^3 - 3 \times 0^2 + 4 = 4$ ✓

Intercept is $(0, 4)$ ✓

Sketch: ✓

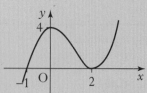

x-intercepts
- You should write down the coordinates of both x-axis intercepts i.e. $(-1, 0)$ and $(2, 0)$.

y-intercept
- The y-axis intercept is expected to be clearly indicated when sketching. It is good practice, as with the x-intercepts, to write down the coordinates, i.e. $(0, 4)$.

Sketch
- Your sketch should clearly show the main features: shape and intercepts and stationary points.

HMRN: p 21

3 marks

22.

For points of intersection

Solve $f(x) = g(x)$

$\Rightarrow 2x^3 + 3x + 12 = 2 + 16x^2 - x^3$ ✓

$\Rightarrow 3x^3 - 16x^2 + 3x + 10 = 0$ ✓

$$\begin{array}{r|rrrr} 1 & 3 & -16 & 3 & 10 \\ & & 3 & -13 & -10 \\ \hline & 3 & -13 & -10 & 0 \end{array}$$ ✓

0 remainder $\Rightarrow 1$ is a root

So $x - 1$ is a factor ✓

The equation becomes:

$(x - 1)(3x^2 - 13x - 10) = 0$ ✓

$\Rightarrow (x - 1)(3x + 2)(x - 5) = 0$ ✓

$\Rightarrow x - 1 = 0$ or $3x + 2 = 0$ ✓

or $x - 5 = 0$

$x = 1$ or $x = -\dfrac{2}{3}$ or $x = 5$

So the x-coordinates are:

For point P: $-\dfrac{2}{3}$

For point Q: 1

For point R: 5 ✓

Strategy
- Setting the two formulae equal to each other and solving is the strategy in this question.

Cubic equation
- Recognising a cubic and rearranging terms in order gains a mark.

Strategy
- Trying particular values of x. In this case $x = 1$ was successful.

Interpretation
- $x = 1$ is a solution of the equation so $x - 1$ is a factor

Factorising
- 3 marks are allocated to this:

1 mark: starting to factorise
$(x - 1)(3x^2 ...)$

1 mark: quadratic factor $3x^2 - 13x - 10$

1 mark: completing the factorisation
$(x - 1)(3x + 2)(x - 5)$

Interpretation
- From the diagram you can assign the 3 values to the points P, Q and R.

HMRN: p 26

8 marks

23. (a)

Upland: $u_{n+1} = 0.35u_n + 2000$ ✓

Lowland:

$v_{n+1} = 0.25\,v_n + 2500$ ✓

2 marks

Recurrence Relations
- Remember to use the "percentage left" in the recurrence relation, e.g. 65% is lost so 35% remains – so use 0.35 not 0.65.

HMRN: p 24

23. (b)

Both recurrence relations have a limit since their multipliers (0.35 and 0.25) lie between -1 and 1. ✓
For Upland let the limit be L. ✓
Then $L = 0.35L + 2000$
$\Rightarrow L - 0.35L = 2000$
$\Rightarrow 0.65L = 2000$

$\Rightarrow L = \dfrac{2000}{0.65} = 3076.9\ldots$ ✓

For Lowland let the limit be M.
Then $M = 0.25M + 2500$
$\Rightarrow M - 0.25M = 2500$
$\Rightarrow 0.75M = 2500$

$\Rightarrow M = \dfrac{2500}{0.75} = 3333.3\ldots$

In the long run the Upland site requires the smaller tank (3077 litres) compared to the Lowland site (3334 litres). ✓

4 marks

Limit Condition
- 1 mark is allocated to a clear statement justifying the use of a limit. There is a limit only if the multiplier lies between -1 and 1

Strategy
- To find the limit: if you apply the recurrence relation you produce the same result,

  e.g. $L = 0.35L + 2000$

Limit
- Calculation of a correct limit gains 1 mark.

Decision
- For the final mark you must correctly calculate the other limit AND clearly make your decision based on a comparison.

HMRN: p 24

24.

$$\cos\theta(\cos\theta - 1) = \sin^2\theta$$

$$\Rightarrow \cos^2\theta - \cos\theta = \sin^2\theta$$

$$\Rightarrow \cos^2\theta - \cos\theta = 1 - \cos^2\theta \quad \checkmark$$

$$\Rightarrow 2\cos^2\theta - \cos\theta - 1 = 0 \quad \checkmark$$

$$\Rightarrow (2\cos\theta + 1)(\cos\theta - 1) = 0 \quad \checkmark$$

$$\Rightarrow 2\cos\theta + 1 = 0 \text{ or } \cos\theta - 1 = 0$$

$$\Rightarrow \cos\theta = -\frac{1}{2} \text{ or } \cos\theta = 1 \quad \checkmark$$

For $\cos\theta = -\frac{1}{2}$:

θ is in 2nd or 3rd quadrants ($-$ve)

1st quadrant angle is $\frac{\pi}{3}$

So $\theta = \pi - \frac{\pi}{3}$ or $\pi + \frac{\pi}{3}$

$\quad = \frac{3\pi}{3} - \frac{\pi}{3}$ or $\frac{3\pi}{3} + \frac{\pi}{3}$

$\quad = \frac{2\pi}{3}$ or $\frac{4\pi}{3}$

For $\cos\theta = 1$:
Use the graph
This gives
$\theta = 0$ or 2π.
In general the solutions are

$\ldots, 0, \frac{2\pi}{3}, \frac{4\pi}{3}, 2\pi, \ldots$

however in this case $\pi < \theta < 2\pi$

So $\theta = \frac{4\pi}{3}$ is the only solution. $\checkmark$

5 marks

Strategy
- Looking at the 'make-up' of the equation will lead you to produce an equation in $\cos\theta$ – so use $\sin^2\theta + \cos^2\theta = 1$ to replace $\sin^2\theta$ by $1 - \cos^2\theta$.

'Standard form'
- The equation is a 'quadratic' in $\cos\theta$ and should be arranged in the standard quadratic form: $ax^2 + bx + c = 0$ in this case $a\cos^2\theta + b\cos\theta + c = 0$.

Factorisation
- $2x^2 - x - 1 = (2x + 1)(x - 1)$ so likewise:
$2\cos^2\theta - \cos\theta - 1 = (2\cos\theta + 1)(\cos\theta - 1)$

Solve for $\cos\theta$
- In this case the 'roots' of the quadratic equation are $-\frac{1}{2}$ and 1. These are the possible values for $\cos\theta$.

Solve for θ
- In general for values of $\sin\theta$ or $\cos\theta$ of -1, 0 or 1 you should use the sine or cosine graph to determine the angles. In this case neither value (0 or 2π) are in the allowed interval ($\pi < \theta < 2\pi$).

- For $\cos\theta = -\frac{1}{2}$ the 2nd quadrant solution is not in the allowed interval ($\pi < \theta < 2\pi$).

HMRN: p 37

WORKED ANSWERS: EXAM A **PAPER 2**

1. (*a*)

$$y = x^2 - \frac{1}{3}x^3 \Rightarrow \frac{dy}{dx} = 2x - x^2 \quad \checkmark$$

At M (3,0) $x = 3$ $\checkmark$

So $\dfrac{dy}{dx} = 2 \times 3 - 3^2 = -3$ $\checkmark$

gradient of tangent is -3.
Point on tangent is (3,0).
Equation of tangent is:
$y - 0 = -3(x - 3)$
$\Rightarrow y = -3x + 9.$ $\checkmark$

4 marks

Strategy
• Knowing to differentiate gains you the 1st mark.

Differentiate
• The correct result gains the 2nd mark.

Calculation
• Where are you on the curve? At the place where $x = 3$, so use $x = 3$ and the gradient formula.

Equation
• Using $y - b = m(x - a)$. In this case $m = -3$ and (a, b) is (3,0).

HMRN: p 20

Rearrangement
• The form $y = -3x + 9$ is necessary for the subsequent substitution. Do not use $x = -\frac{1}{3}y + 3$ as this involves fraction work leading to errors.

1. (*b*)
To find the points of intersection of line and circle
Solve: $\left. \begin{array}{l} y = -3x + 9 \\ x^2 + y^2 - 4x - 26y + 163 = 0 \end{array} \right\}$ $\checkmark$

Substitute $y = -3x + 9$ in the circle equation: $\checkmark$
$x^2 + (-3x + 9)^2 - 4x -$
$26(-3x + 9) + 163 = 0$
$\Rightarrow x^2 + 9x^2 - 54x + 81 - 4x + 78x$
$\quad - 234 + 163 = 0$
$\Rightarrow 10x^2 + 20x + 10 = 0$ $\checkmark$
$\Rightarrow 10(x^2 + 2x + 1) = 0$
$\Rightarrow 10(x + 1)(x + 1) = 0$
$\Rightarrow x = -1$ $\checkmark$
Since there is only one solution the line is a tangent to the circle. $\checkmark$
when $x = -1$ $y = -3 \times (-1) + 9 = 12$
So N(−1,12) is the point of contact. $\checkmark$

6 marks

Strategy
• Evidence of substitution gains you the strategy mark.

'Standard form'
• This is a quadratic equation and should be written in the 'standard' way, i.e. $ax^2 + bx + c = 0$.

Solution
• Notice that removing a common factor reduces the magnitude of the coefficients and makes the rest of the factorisation easier.

Proof
• A clear statement is required — there was one point of intersection so you have a tangent.

Calculation
• A point was asked for so you must give the coordinates, $x = -1$ and $y = 12$ is not enough.

HMRN: p 40

1. (c)

when $x = 0$ $y = -3 \times 0 + 9 = 9$ ✓

The required ✓
ratio is 3:1 ✓

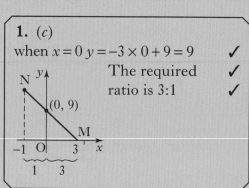

3 marks

Strategy
• Knowing where the y-intercept is gains you this strategy mark.

Strategy
• An alternative approach uses vectors. If y-intercept is P.
$$\overrightarrow{MP} = \begin{pmatrix} -3 \\ 9 \end{pmatrix} \text{ and } \overrightarrow{PN} = \begin{pmatrix} -1 \\ 3 \end{pmatrix} \text{ so } \overrightarrow{MP} = 3\overrightarrow{PN}$$
etc.

Ratio
• Deducing the correct ratio gains you this final mark. Note 1:3 will not gain this mark.

HMRN: p 44–45

2. (a)

$$M\left(\frac{-3+(-3)}{2}, \frac{4+(-4)}{2}\right) = M(-3, 0) ✓$$

For M(−3,0) and R(5,12)
$$m_{MR} = \frac{12-0}{5-(-3)} = \frac{12}{8} = \frac{3}{2}$$ ✓

gradient of median is $\frac{3}{2}$

point on median is $(-3, 0)$

so equation is $y - 0 = \frac{3}{2}(x - (-3))$
$\Rightarrow 2y = 3(x + 3) \Rightarrow 2y = 3x + 9$
$\Rightarrow 2y - 3x = 9$ ✓

3 marks

Interpretation
• Did you know what a median is? The line from a vertex to the midpoint of the opposite side.

Gradient
• The formula used here is:
$$A(x_1, y_1), B(x_2, y_2)$$
$$m_{AB} = \frac{y_2 - y_1}{x_2 - x_1}$$

• Perpendicular gradients are not required for medians, only for altitudes.

Equation
• Using $y - b = m(x - a)$ with $m = \frac{3}{2}$ and (a, b) being $(-3, 0)$

HMRN: p 7

2. (b)
For P(−3,−4) and R(5,12) ✓

$$m_{PR} = \frac{12-(-4)}{5-(-3)} = \frac{16}{8} = 2$$

$$\Rightarrow m_\perp = -\frac{1}{2}$$ ✓

gradient of altitude is $-\frac{1}{2}$

point on altitude is Q(−3,4)

equation is $y - 4 = -\frac{1}{2}(x - (-3))$
$\Rightarrow 2y - 8 = -(x + 3)$
$\Rightarrow 2y - 8 = -x - 3$
$\Rightarrow 2y + x = 5$ ✓

3 marks

Strategy
• Finding the gradient of the "base" PR is the essential 1st step here.

Perpendicular gradient
• Did you know what an altitude is? A line from a vertex perpendicular to the opposite side.

• The result used is $m = \frac{a}{b} \Rightarrow m_\perp = -\frac{b}{a}$
invert and change the sign.

Equation
• Using $y - b = m(x - a)$ with $m = -\frac{1}{2}$ and (a, b) being the point $(-3,4)$.

HMRN: p 7

2. (*c*)
To find the point of intersection:
Solve $\left.\begin{array}{l} 2y + x = 5 \\ 2y - 3x = 9 \end{array}\right\}$ ✓

Subtract: $4x = -4 \Rightarrow x = -1$ ✓
Substitute x = –1 in $2y + x = 5$
$\quad \Rightarrow 2y - 1 = 5 \Rightarrow 2y = 6$
$\quad \Rightarrow y = 3$
The point of intersection is
$\quad S(-1, 3)$ ✓

3 marks

Strategy
• Simultaneous equations: if your method is clear you will gain this strategy mark.

Calculation
• Correct calculation of either x or y gains the 2nd mark.

Calculation
• Correct calculation of the other variable gains you this last mark.

HMRN: p 6

2. (*d*)
from (*b*) $m_{PR} = 2$
For S(–1,3) and T(2,9)
$m_{ST} = \dfrac{9 - 3}{2 - (-1)} = \dfrac{6}{3} = 2$ ✓
So $m_{ST} = m_{PR} = 2$
and so ST is parallel to PR. ✓

2 marks

Gradient
• You must find the gradient of this new line ST if you are to compare its slope with that of line PR.

Parallel lines
• If the gradients of two lines are equal then the lines are parallel.

• Clear statements are needed here so that it is obvious you understand the result:

$\quad$ equal gradients $\Leftrightarrow$ parallel lines.

HMRN: p 4

3. (*a*)
$\quad$ PQ = 4 units
$\quad$ QR = 5 units
$\quad$ RV = 6 units ✓

1 mark

Interpretation
• The difficulty is that the axes are not shown. Look for a single change in coordinates: P(–1,–1,–1) to S (–1,4,–1): This is 5 units parallel to the y-axis since only the y-coordinate has changed.

HMRN: p 42

3. (b)

P(−1, −1, −1)   V(3, 4, 5)

S(−1, 4, −1)

$\overrightarrow{VS} = s - v$

$$= \begin{pmatrix} -1 \\ 4 \\ -1 \end{pmatrix} - \begin{pmatrix} 3 \\ 4 \\ 5 \end{pmatrix} = \begin{pmatrix} -4 \\ 0 \\ -6 \end{pmatrix}$$

$\overrightarrow{VP} = p - v = \begin{pmatrix} -1 \\ -1 \\ -1 \end{pmatrix} - \begin{pmatrix} 3 \\ 4 \\ 5 \end{pmatrix} = \begin{pmatrix} -4 \\ -5 \\ -6 \end{pmatrix}$ ✓

$\overrightarrow{VS} . \overrightarrow{VP} = \begin{pmatrix} -4 \\ 0 \\ -6 \end{pmatrix} . \begin{pmatrix} -4 \\ -5 \\ -6 \end{pmatrix}$ ✓

$= -4 \times (-4) + 0 \times (-5) + (-6) \times (-6)$

$= 52$

$|\overrightarrow{VS}| = \left| \begin{pmatrix} -4 \\ 0 \\ -6 \end{pmatrix} \right| = \sqrt{(-4)^2 + 0^2 + (-6)^2}$ ✓

$\quad = \sqrt{16 + 0 + 36} = \sqrt{52}$

$|\overrightarrow{VP}| = \left| \begin{pmatrix} -4 \\ -5 \\ -6 \end{pmatrix} \right| = \sqrt{(-4)^2 + (-5)^2 + (-6)^2}$ ✓

$\quad = \sqrt{16 + 25 + 36} = \sqrt{77}$

so $\cos P\hat{V}S = \dfrac{\overrightarrow{VS} . \overrightarrow{VP}}{|\overrightarrow{VS}||\overrightarrow{VP}|} = \dfrac{52}{\sqrt{52}\sqrt{77}}$ ✓ ✓

$\Rightarrow P\hat{V}S = \cos^{-1}\left(\dfrac{52}{\sqrt{52}\sqrt{77}} \right) = 34 \cdot 73 ... °$

angle PVS = $34 \cdot 7°$ (to 1
decimal place) ✓

7 marks

Components

- The result $\overrightarrow{AB} = b - a$ where a is the position vector of A and b is the position vector of B is used frequently to find components.

- Notice that the vector arrows point outwards from the vertex of the angle. You must always do this when calculating the angle between vectors.

Dot product

- The result is $\begin{pmatrix} x_1 \\ y_1 \\ z_1 \end{pmatrix} . \begin{pmatrix} x_2 \\ y_2 \\ z_2 \end{pmatrix} = x_1 x_2 + y_1 y_2 + z_1 z_2$

Lengths

- The result used is: $\left| \begin{pmatrix} a \\ b \\ c \end{pmatrix} \right| = \sqrt{a^2 + b^2 + c^2}$

- Careful with negatives. Squaring always produces a positive or zero quantity – never a negative quantity.

Angle formula

- The result used is: $\cos\theta = \dfrac{a.b}{|a||b|}$

Calculation

- On the calculator:

$$\cos^{-1}\left(52 \div (\sqrt{52} \times \sqrt{77}) \right)$$

- Radians acceptable: $0 \cdot 606$

HMRN: p 46

4. (*a*)

$3\cos x° - 2\sin x° = k\cos(x + a)°$ ✓

$\Rightarrow 3\cos x° - 2\sin x°$

$= k\cos x° \cos a° - k\sin x° \sin a°$

now compare coefficients of

$\cos x°$ and $\sin x°$:   ✓

$\left.\begin{array}{l} k\cos a° = 3 \\ k\sin a° = 2 \end{array}\right\}$ since $\cos a° > 0$ and $\sin a° > 0$ $a°$ is in 1st quadrant.

Divide : $= \dfrac{k\sin a°}{k\cos a°} = \dfrac{2}{3}$

$\Rightarrow \tan a° = \dfrac{2}{3}$ ✓

$\Rightarrow a \doteq 33 \cdot 7$

Square and Add:

$(k\cos a°)^2 + (k\sin a°)^2 = 3^2 + 2^2$

$\Rightarrow k^2 \cos^2 a° + k^2 \sin^2 a° = 9 + 4$

$\Rightarrow k^2(\cos^2 a° + \sin^2 a°) = 13$

$\Rightarrow k^2 \times 1 = 13 \Rightarrow k = \sqrt{13}\ (k > 0)$ ✓

so $3\cos x° - 2\sin x°$

$\quad = \sqrt{13}\cos(x + 33 \cdot 7)°$

4 marks

Addition Formula

- Expanding $k\cos(x + a)°$ is the essential first step here. The formula:

$$\cos(A \pm B) = \cos A \cos B \mp \sin A \sin B$$

is on the formula sheet given in the exam.

Coefficients

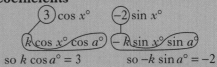

so $k\cos a° = 3$ so $-k\sin a° = -2$

Calculation

- You are using the result $\frac{\sin a°}{\cos a°} = \tan a°$ and cancellation of the "k"s to find the angle $a°$.

Calculation

- The result $\sin^2 a° + \cos^2 a° = 1$ ensures the disappearance of $a°$ and allows the calculation of k. k is always positive.

HMRN: p 53–54.

4. (b)

$3 \cos x° - 2 \sin x° = 2$

$\Rightarrow \sqrt{13} \cos(x + 33·7)° = 2$ ✓

$\Rightarrow \cos(x + 33·7)° = \dfrac{2}{\sqrt{13}}$

$x + 33·7°$ is in 1st or 4th quadrant (positive)

1st quadrant: $\cos^{-1}\left(\dfrac{2}{\sqrt{13}}\right) \doteqdot 56·3°$

$\Rightarrow x + 33·7 = 56·3$

$\Rightarrow x + 56·3 - 33·7 = 22·6$ ✓

4th quadrant:

$x + 33·7 = 360 - 56·3 = 303·7$

$\Rightarrow x = 303·7 - 33·7 = 270·0$ ✓

Solution is

$x = 22·6$ or $270·0$

(to 1 decimal place)

3 marks

Set up equation

• Using the result you obtained in part (a) the equation can be rewritten.

• The aim is to change more complicated equations into the format:

$$\cos(\text{angle}) = \text{number}$$

in this case angle $= x + 33·7$

and number $= \dfrac{2}{\sqrt{13}}$

1st solution

• The 1st quadrant solution is $56·3°$. In other words the angle $= 56·3° \Rightarrow x + 33·7 = 56·3°$. But solving the original equation means you have to find the value of x not $x + 33·7$ so there is a further step: $x = 56·3 - 33·7$.

2nd solution

• Cosine is positive in the 1st and 4th quadrants.

S	A˘
T	C˘

HMRN: p 54

---

5. (a)

$P(h) = P_0 = e^{-kh}$

In this case $P(4·95) = \dfrac{1}{2}P_0$ ✓

$\Rightarrow P_0 e^{-k \times 4·95} = \dfrac{1}{2}P_0 \Rightarrow e^{-4·95k} = \dfrac{1}{2}$

$\Rightarrow \log_e \dfrac{1}{2} = -4·95k \Rightarrow k = \dfrac{\log_e \frac{1}{2}}{-4·95}$ ✓

$\Rightarrow k = 0·1400$ (to 4 decimal places) ✓

3 marks

Interpretation

• You are finding k given that you know $P(h)$ and you know h

$$P(h) \quad = \quad P_0 e^{-kh}$$

This is $\frac{1}{2}P_0$ This is $4·95$

Log statement

• Using the result: $b^c = a \leftrightarrow \log_b a = c$

$e^{-4·95k} = \frac{1}{2}$ becomes $\log_e \frac{1}{2} = -4·95k$

Calculation

• Remember the $\boxed{\ln}$ button for "$\log_e$".

HMRN: p 50–51

5. (b)

The formula is $P(h) = P_0 e^{-0.14h}$

For top of Everest $h = 8.85$

$P(8.85) = P_0 e^{-0.14 \times 8.85}$

$\qquad = P_0 \times 0.2896\ldots$ ✓

This is a reduction of

$100\% - 28.96\ldots\%$

$= 71.03\ldots\% \doteq 71\%$ ✓

2 marks

Substitution
- Careful with the units. In the original formula h is in kilometres so 8.85 is used, not 8850.

Calculation
- $\boxed{e^x}$ button is used for $e^{-0.14 \times 8.85}$ so:

$\boxed{e^x}\,\boxed{(}\,\boxed{(}\,\boxed{(-)}\,\boxed{0}\,\boxed{\cdot}\,\boxed{1}\,\boxed{4}\,\boxed{\times}\,\boxed{8}\,\boxed{\cdot}\,\boxed{8}\,\boxed{5}\,\boxed{)}\,\boxed{\text{EXE}}$

or the equivalent on your calculator!

HMRN: p 50–51

6. (a)

To find the points of intersection

Solve: $\left.\begin{array}{l} y = x + 2 \\ y = 6 + 4x - x^2 \end{array}\right\}$

so $x + 2 = 6 + 4x - x^2$ ✓

$\Rightarrow x^2 - 3x - 4 = 0$

$\Rightarrow (x + 1)(x - 4) = 0$

$\Rightarrow x = -1$ or $x = 4$ ✓

Area =

$\int_{-1}^{4} (6 + 4x - x^2) - (x + 2)\, dx$

$= \int_{-1}^{4} 6 + 4x - x^2 - x - 2\, dx$ ✓

$= \int_{-1}^{4} 4 + 3x - x^2\, dx = \left[4x + \frac{3x^2}{2} - \frac{x^3}{3} \right]_{-1}^{4}$ ✓

$= \left(4 \times 4 + \frac{3 \times 4^2}{2} - \frac{4^3}{3} \right)$

$\quad - \left(4 \times (-1) + \frac{3 \times (-1)^2}{2} - \frac{(-1)^3}{3} \right)$ ✓

$= 16 + 24 - \frac{64}{3} + 4 - \frac{3}{2} - \frac{1}{3}$ ✓

$= 44 - \frac{65}{3} - \frac{3}{2} = \frac{264}{6} - \frac{130}{6} - \frac{9}{6}$

$= \frac{264 - 130 - 9}{6} = \frac{125}{6}$

Enclosed area $= \frac{125}{6}$ unit^2 ✓

7 marks

Strategy
- To find the area you will be integrating so will need to know the limits. Hence the need to set the equations equal to each other to find the points of intersection. The x values for these points are the limits used in the integration.

Solve equation
- Never attempt to arrange a quadratic equation with a negative coefficient for the x^2 term – the factorising is then more difficult.

Strategy
- This mark is for knowing how to find the enclosed area: $\int$(top curve) minus (bottom curve)

Limits
- Work left to right on diagram (-1 to 4) and bottom to top on integral: $\int_{-1}^{4}$

Integration
- Here you are using. $\int ax^n dx = \frac{ax^{n+1}}{n+1}$.

No constant is needed when there are limits on the integral sign.

Substitution
- Careful with the order:

$\left(\begin{array}{c} x = 4 \\ \text{Substitution} \end{array} \right) - \left(\begin{array}{c} x = -1 \\ \text{Substitution} \end{array} \right)$ using the result $\int_a^b f(x)\, dx = F(b) - F(a)$ where $F(x)$ is the result of integrating $f(x)$.

Calculation
- Take great care, even with a calculator, as these are not easy calculations!
- Always give exact answers.

HMRN: p 33

6. (b)

$$A(x) = -\frac{5}{2}x^2 + \frac{15}{2}x + 10$$

$$\Rightarrow A'(x) = -5x + \frac{15}{2}$$

For stationary value set $A'(x) = 0$ ✓

$$\Rightarrow -5x + \frac{15}{2} = 0 \Rightarrow -5x = -\frac{15}{2} \Rightarrow x = \frac{3}{2}.$$

x:	$\frac{3}{2}$	
$A'(x)$:	+	−
Shape of graph:	╱	╲
nature:	max	

So $x = \frac{3}{2}$ gives a maximum value

$$A\left(\frac{3}{2}\right) = -\frac{5}{2} \times \left(\frac{3}{2}\right)^2 + \frac{15}{2} \times \frac{3}{2} + 10$$

$$= -\frac{45}{8} + \frac{45}{4} + 10 = \frac{-45}{8} + \frac{90}{8} + \frac{80}{8}$$

$$= \frac{-45 + 90 + 80}{8} = \frac{125}{8}$$

Required fraction $= \dfrac{\frac{125}{8}}{\frac{125}{6}} = \dfrac{6}{8} = \dfrac{3}{4}.$ ✓

4 marks

Strategy
- For a maximum value you must find a stationary point where the gradient on the graph is zero, hence set $A'(x) = 0$.

Differentiate and solve
- One method of solving the equation is:

$$-5x(\times 2) = -\frac{15}{2}(\times 2) \Rightarrow -10x = -15$$

$$\Rightarrow x = \frac{-15}{-10}$$

So $x = \frac{3}{2}$

Justify
- You must show $x = \frac{3}{2}$ gives a maximum value hence the need for the 'nature table'.

Interpretation
- You have found that when $x = \frac{3}{2}$ the area of the shaded triangle is at a maximum. The actual shaded area is given by $A\left(\frac{3}{2}\right)$.

Your calculation should give $\frac{125}{8}$ for this area. In part (a) the whole enclosed area was found to be $\frac{125}{6}$ unit^2. To calculate the fraction consider a simpler case:

What fraction is 2 unit^2 of 6 unit^2? It's $\frac{2}{6}$ or $\frac{1}{3}$

What fraction is

$\frac{125}{8}$ unit^2 of $\frac{125}{6}$ unit^2? It's $\dfrac{\frac{125}{8}}{\frac{125}{6}}$

HMRN: p 20–21

7.

In triangle PQR

$RQ^2 = PR^2 + PQ^2$

$\qquad = 5^2 + 12^2 = 25+144 = 169$ ✓

So $RQ = \sqrt{169} = 13$ ✓

$\Rightarrow \cos 2\theta = \dfrac{5}{13}$ ✓

$\Rightarrow 2\cos^2 \theta - 1 = \dfrac{5}{13}$

$\Rightarrow 2\cos^2 \theta = \dfrac{5}{13} + 1 = \dfrac{18}{13}$

$\Rightarrow \cos^2 \theta = \dfrac{9}{13}$

$\Rightarrow \cos\theta = \pm\sqrt{\dfrac{9}{13}}$

$\qquad = \pm\dfrac{3}{\sqrt{13}}$ but $0 < \theta < \dfrac{\pi}{2}$ ✓

so $\cos\theta = \dfrac{3}{\sqrt{13}} = \dfrac{3\times\sqrt{13}}{\sqrt{13}\times\sqrt{13}}$ ✓

$\qquad = \dfrac{3\sqrt{13}}{13}$.

5 marks

Strategy
- You are using SOHCAHTOA in the large right-angled triangle PQR with angle 2θ.

Value
- Pythagoras' Theorem allows you to calculate the length of the hypotenuse RQ then give the value of $\cos 2\theta$ as $\dfrac{5}{13}$.

Strategy
- The Double angle formula allows you to calculate the value of $\cos\theta$.

Value
- 1st quadrant so the value of $\cos\theta$ is positive $\dfrac{3}{\sqrt{13}}$.

Rationalisation
- Rationalising the denominator – show this clearly.

HMRN: p 35–36

1.

$$x^2 - x - 2 = 0$$

Compare $ax^2 + bx + c = 0$

$a = 1$, $b = -1$, $c = -2$

Discriminant $= b^2 - 4ac$

$$= (-1)^2 - 4 \times 1 \times (-2)$$

$$= 1 + 8 = 9 > 0$$

So there are two distinct Real roots.

Also $9 = 3^2$, a perfect square.

So the roots are rational.

Choice D

2 marks

- The 1st result you are using is concerning the value of the discriminant $b^2 - 4ac$. To solve $ax^2 + bx + c = 0$ the roots are

$$x = \frac{-b + \sqrt{b^2 - 4ac}}{2a} \text{ and } x = \frac{-b - \sqrt{b^2 - 4ac}}{2a}$$

If $b^2 - 4ac > 0$ as is the case in this question there will be two distinct Real values for these roots. If $b^2 - 4ac = 0$ the two values above are equal (one root). If $b^2 - 4ac < 0$ then $\sqrt{b^2 - 4ac}$ is not Real (no roots).

- $\sqrt{9} = 3$ and so no surd remains in the above expressions. The roots are rational.

HMRN: p 27

2.

$u_{n+1} = 0{\cdot}8u_n + 3$, $u_0 = 5$

So $u_1 = 0{\cdot}8 \times 5 + 3 = 7$

and $u_2 = 0{\cdot}8 \times 7 + 3 = 8{\cdot}6$

Choice B

2 marks

- The process is as follows:

$u_0 \longrightarrow u_1 \longrightarrow u_2$

multiply by   multiply by
   0·8 then   0·8 then
    add 3   add 3
giving:
    5 $\longrightarrow$ 7 $\longrightarrow$ 8·6

HMRN: p 23

3.

$m_{AB} = \tan 40°$

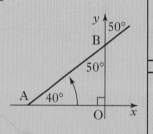

Choice D

2 marks

- The result used here is gradient $m = \tan \theta°$

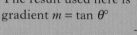

- The angle used is the angle the line makes with the positive direction of the x-axis. The 50° given in the question is the angle with the y-axis - that's the wrong axis!

HMRN: p 4

4.

Point of contact is $(0,5)$.

Centre is $(0,3)$

So radius = 2 units

Centre $(0,3)$ radius = 2

So equation of circle is :

$(x-0)^2 + (y-3)^2 = 2^2$

$\Rightarrow x^2 + (y-3)^2 = 4$

Choice A

2 marks

- To find the equation of a circle you need to know two facts:

  The centre: (a, b) The radius: r

  The equation is then:

  $(x-a)^2 + (y-b)^2 = r^2$

- In this case the centre is known, $(0,3)$, so what is the radius? The realisation is that the radius from the centre to the point of contact with the tangent lies on the y-axis and so the difference in height, i.e. $5-3=2$ is the length of the radius.

  HMRN: p 39

5.

$\sqrt{2^2+1^2}$
$= \sqrt{5}$, 1

p°

2

$\tan p^\circ = \dfrac{1}{2}$

so $\cos(p-q)^\circ$

$= \cos p^\circ \cos q^\circ + \sin p^\circ \sin q^\circ$

$= \dfrac{2}{\sqrt{5}} \times \cos q^\circ + \dfrac{1}{\sqrt{5}} \times \sin q^\circ$

Choice D

2 marks

- When you are given $\sin$, $\cos$ or $\tan$ equal to a number that is not an 'exact value' like $\dfrac{1}{2}, \dfrac{\sqrt{3}}{2}, \dfrac{1}{\sqrt{2}}$ etc. then you should draw a right-angled triangle and use Pythagoras' Theorem.

- The addition formula used here is given to you in the exam:

  $\cos(A \pm B) = \cos A \cos B \mp \sin A \sin B$

  – notice in the cosine addition formula the signs change: + becomes – and – becomes +.

  HMRN: p 35

6.

p and q are perpendicular

$\Rightarrow p.q = 0$

$\Rightarrow \begin{pmatrix} a \\ -1 \\ 2 \end{pmatrix} \cdot \begin{pmatrix} -1 \\ a \\ 3 \end{pmatrix} = 0$

$\Rightarrow a \times (-1) + (-1) \times a + 2 \times 3 = 0$

$\Rightarrow -a + (-a) + 6 = 0$

$\Rightarrow -2a + 6 = 0 \Rightarrow 2a = 6 \Rightarrow a = 3$

Choice D

2 marks

- The result used here is:

  p and q are perpendicular vectors $\Leftrightarrow p.q = 0$

  (p and q are non-zero vectors)

- Remember that

  $\cos \theta^\circ = \dfrac{p.q}{|p||q|}$

  p and q perpendicular means $\theta^\circ = 90^\circ$ so $\cos \theta^\circ = \cos 90^\circ = 0$. If the fraction on the right is zero then the numerator $p.q$ is zero.

  HMRN: p 46

7.

$y = f(x) \rightarrow y = -f(x)$
This is reflection
in the x-axis

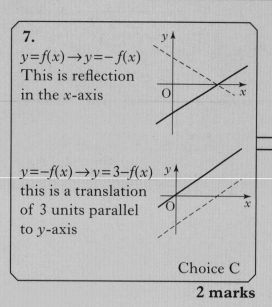

$y = -f(x) \rightarrow y = 3 - f(x)$
this is a translation
of 3 units parallel
to y-axis

Choice C

2 marks

- One way of checking your choice of answer is to work with the point $(0,3)$. The equation is $y = f(x)$ so if $x = 0$ this gives $y = 3$. This means $3 = f(0)$.

  So what happens in the new equation when $x = 0$? $y = 3 - f(x)$ gives $y = 3 - f(0) = 3 - 3 = 0$.

  So the origin lies on the new line. Only choice C and D show this. Knowing $-f(x)$ is a 'flip' in the x-axis means a downward slope will change to an upward slope – choice C.

  HMRN: p 9

8.

Let the limit be L

$\Rightarrow L = 0{\cdot}9L + 90$

$\Rightarrow L - 0{\cdot}9L = 90$

$\Rightarrow 0{\cdot}1L = 90 \Rightarrow L = \dfrac{90\,(\times 10)}{0{\cdot}1\,(\times 10)}$

$\Rightarrow L = \dfrac{900}{1} = 900$

Choice D

2 marks

- The multiplier here is $0{\cdot}9$ so you know a limit exists as this number lies between -1 and 1.

- The formula $L = \dfrac{b}{1-a}$ is not given to you in the exam and many mistakes arise from muddling 'a' and 'b'. It is recommended that you understand the algebraic method given in the solution:

  apply the recurrence relation to L and you will still get L as L is the limit.

  HMRN: p 24

9.

$C(0, -3)$ and $P(-2, 0)$

$m_{PC} = \dfrac{-3 - 0}{0 - (-2)} = \dfrac{-3}{2} = -\dfrac{3}{2}$

$\Rightarrow m_{\perp} = \dfrac{2}{3}$. So for the tangent:

A point is $P(-2, 0)$ and the

gradient $= \dfrac{2}{3}$

Equation is: $y - 0 = \dfrac{2}{3}(x - (-2))$

$\Rightarrow y = \dfrac{2}{3}(x + 2)$

Choice A

2 marks

- The tangent is perpendicular to the radius from the centre C to the point of contact P.

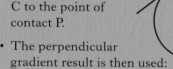

- The perpendicular gradient result is then used:

  if $m = \dfrac{a}{b}$ then $m_{\perp} = -\dfrac{b}{a}$

  (invert and change the sign)

- Use $y - b = m(x - a)$ with $m = \dfrac{2}{3}$ and (a, b) being the point $P(-2, 0)$.

  HMRN: p 4

10.

$\sqrt{2}\cos\theta + 1 = 0 \Rightarrow \sqrt{2}\cos\theta = -1$

$\Rightarrow \cos\theta = -\dfrac{1}{\sqrt{2}}$. Required solution is in the 3rd quadrant.

1st quadrant angle $= \pi/4$

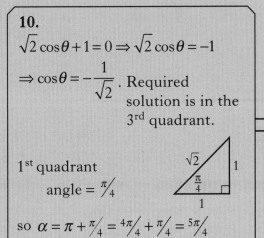

so $\alpha = \pi + \pi/4 = {}^{4\pi}/_4 + \pi/4 = {}^{5\pi}/_4$

Choice B

2 marks

- Lots of greek letters here!

  θ is the unknown angle.

  α is the particular value of θ that you are trying to find, i.e the root or solution of the equation.

  $\pi \leq \alpha \leq {}^{3\pi}/_2$ indicates that your solution is in the 3rd quadrant.

- You should recognise $\dfrac{1}{\sqrt{2}}$ as an 'exact value'.

  HMRN: p 34

11.

$\displaystyle\int 6\cos 2x\, dx$

$= \dfrac{6\sin 2x}{2} + c$

$= 3\sin 2x + c$

Choice B

2 marks

- The result used is:

  $$\int \cos ax\, dx = \frac{\sin ax}{a} + c$$

- In your exam you will be given this result in this format:

$f(x)$	$\int f(x)\,dx$
$\cos ax$	$\frac{1}{a}\sin ax + c$

  HMRN: p 49

12.

$f(x) = \sqrt{x^2 + 1} = (x^2 + 1)^{\frac{1}{2}}$

$\Rightarrow f'(x) = \dfrac{1}{2}(x^2 + 1)^{-\frac{1}{2}} \times 2x$

$= x(x^2 + 1)^{-\frac{1}{2}}$

$= \dfrac{x}{(x^2+1)^{\frac{1}{2}}} = \dfrac{x}{\sqrt{x^2+1}}$

Choice B

2 marks

- This is the use of the 'chain rule'. The basic result is:

  $$f(x) = x^{\frac{1}{2}} \Rightarrow f'(x) = \frac{1}{2}x^{-\frac{1}{2}}$$

  If x is replaced by a more complicated expression [let's call it $g(x)$ (in this case $x^2 + 1$)] you use:

  $$f(x) = (g(x))^{\frac{1}{2}} \Rightarrow f'(x) = \frac{1}{2}(g(x))^{-\frac{1}{2}} \times g'(x).$$

  In this case the factor $g'(x)$ is $2x$.

- The other results used are:

  $$\sqrt{a} = a^{\frac{1}{2}} \text{ and } a^{-n} = \frac{1}{a^n}$$

  HMRN: p 48–49

13.

working through choices:

For $\boldsymbol{p}$: $\overrightarrow{BC} = \overrightarrow{BA} + \overrightarrow{AD} + \overrightarrow{DC}$

$$\text{so } \boldsymbol{p} = \boldsymbol{s} - \boldsymbol{u} - \boldsymbol{q}$$

$$= -\boldsymbol{q} + \boldsymbol{s} - \boldsymbol{u}$$

so A is true.

For q : $\overrightarrow{CD} = \overrightarrow{CB} + \overrightarrow{BA} + \overrightarrow{AD}$

$$\text{so } \boldsymbol{q} = -\boldsymbol{p} + \boldsymbol{s} - \boldsymbol{u}$$

so B is false

Choice B

2 marks

- The direction of the arrows is very important. If you travel against the arrow it introduces a negative:

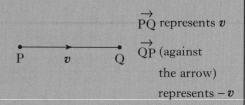

$\overrightarrow{PQ}$ represents $\boldsymbol{v}$

$\overrightarrow{QP}$ (against the arrow) represents $-\boldsymbol{v}$

- In a question where you have to work through the choices you should double check by working through, in this case, Choice C and Choice D – both statements you will find are true.

HMRN: p 43

14.

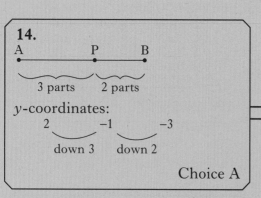

Choice A

2 marks

- The numbers are easy enough to reason out, with -1 as the only suitable choice. Normally the method would be, for example,

$$\overrightarrow{AP} = \frac{3}{5}\overrightarrow{AB}$$

$$\Rightarrow 5\overrightarrow{AP} = 3\overrightarrow{AB} \Rightarrow 5(\boldsymbol{p} - \boldsymbol{a}) = 3(\boldsymbol{b} - \boldsymbol{a})$$

$$\Rightarrow 5\boldsymbol{p} - 5\boldsymbol{a} = 3\boldsymbol{b} - 3\boldsymbol{a} \Rightarrow 5\boldsymbol{p} = 3\boldsymbol{b} + 2\boldsymbol{a}$$

$$= 3\begin{pmatrix} 7 \\ -3 \\ 1 \end{pmatrix} + 2\begin{pmatrix} -3 \\ 2 \\ 6 \end{pmatrix}$$

$$\text{so } 5\boldsymbol{p} = \begin{pmatrix} 15 \\ -5 \\ 15 \end{pmatrix} \Rightarrow \boldsymbol{p} = \begin{pmatrix} 3 \\ -1 \\ 3 \end{pmatrix} \text{ with } -1 \text{ as the}$$

y-coordinate

You probably prefer working through the choices!

HMRN: p 44–45

15.

x-intercept $(-1, 0)$

corresponds to factor $(x + 1)$

x-intercept $(3, 0)$

corresponds to factor $(x - 3)$

so $y = k(x + 1)(x - 3)$

since $(1,8)$ lies on the curve

$y = 8$ when $x = 1$

So $8 = k \times (1 + 1) \times (1 - 3)$

$\Rightarrow 8 = k \times 2 \times (-2) \Rightarrow 8 = -4k$

$\Rightarrow k = \dfrac{8}{-4} = -2$

So $y = -2(x + 1)(x - 3)$

Choice A

2 marks

- Remember that to find x-axis intercepts you set $y = 0$ so for the graph with equation $y = k\,(x - a)(x - b)$ setting $y = 0$ gives:

  $k(x - a)(x - b) = 0 \Rightarrow x - a = 0$ or $x - b = 0$

  $\Rightarrow x = a$ or $x = b$ and the x-intercepts are the points $(a, 0)$ and $(b, 0)$. In the example the intercepts are $(-1, 0)$ and $(3, 0)$ leading to $a = -1$ and $b = 3$ with the factors being $x - a = x - (-1) = x + 1$ and $x - b = x - 3$

- k is the 'scaling factor' in the y axis direction. There are many curves with equation $y = k(x + 1)(x - 3)$:

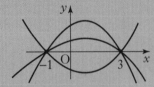

only one passes through $(1,8)$ namely $y = -2(x + 1)(x - 3)$ with $k = -2$.

HMRN: p 29

16.

$y = k \times 2^{-x}$

$(3,1)$ lies on the curve so

$y = 1$ when $x = 3$

$\Rightarrow 1 = k \times 2^{-3} \Rightarrow 1 = k \times \dfrac{1}{2^3}$

$\Rightarrow k = 2^3 = 8$

Choice D

2 marks

- There is an important general result about the graph with equation $y = f(x)$:

  point (a, b) lies on the graph $\Leftrightarrow b = f(a)$

  In other words the values of the coordinates of a point on the graph will satisfy the equation of the graph.

- The result $a^{-n} = \dfrac{1}{a^n}$ is also used in this question.

HMRN: p 50

17.

$3x^2 - 6x + 11$

$= 3\left[x^2 - 2x + \dfrac{11}{3} \right]$

$= 3\left[(x - 1)(x - 1) - 1 + \dfrac{11}{3} \right]$

$= 3\left[(x - 1)^2 + \dfrac{8}{3} \right]$

$= 3(x - 1)^2 + 8$

compare $3(x + a)^2 + b$

$\Rightarrow b = 8$

Choice C

2 marks

- Each term has been divided by 3 but the factor 3 outside the square brackets will eventually multiply each term by 3 so that everything remains the same!

- In these 'completing the square' questions it is the coefficient of x^2 (in this case 3) that causes complications. You should always take this coefficient outside brackets and then divide each term by it. This is what has been done in this example.

- Alternative method in this case is:

  $3[x^2 - 2x] + 11$

  $= 3[(x - 1)(x - 1) - 1] + 11$

  $= 3(x - 1)^2 - 3 + 11$

  avoiding fractions!

HMRN: p 13

18.

$v.(v-w)$

$= v.v - v.w$

$= |v||v|\cos 0° - |v||w|\cos 60°$

$= 3\times 3\times 1 - 3\times 3\times \dfrac{1}{2}$

$= 9 - \dfrac{9}{2} = \dfrac{18}{2} - \dfrac{9}{2} = \dfrac{9}{2}$

Choice B

2 marks

- The following properties of the dot product need to be known for this question:

$$a.(b\pm c) = a.b \pm a.c$$

and $a.b = |a||b|\cos\theta°$

(this last formula is given on your formulae sheet)

- You also need to know:

$$\cos 60° = \frac{1}{2} \text{ and } \cos 0° = 1$$

- $|v|$ and $|w|$ are the magnitudes of the vectors and are represented by the lengths of the lines in the diagram – all 3 units.

HMRN: p 47

19.

Division by zero is not allowed

Consider: $2(x^2 - 3x + 2) = 0$

$\Rightarrow 2(x-1)(x-2) = 0$

$\Rightarrow x = 1 \text{ or } x = 2$

So all Real numbers are allowed apart from $x = 1$ and $x = 2$

Choice C

2 marks

- The 'Domain' of a function f is the set of numbers you are allowed to use in the formula for the function. The most common restrictions arise from two sources:

1. Square roots of negative quantities

2. Division by zero

- In this case the formula is

$$f(x) = \frac{5}{2(x^2 - 3x + 2)}.$$

There are no square roots so the only potential problem is a division by zero. So the question to ask is: can $2(x^2 - 3x + 2)$ ever take the value zero. To answer this involves solving $2(x^2 - 3x + 2) = 0$ as is shown in the solution to the question.

HMRN: p 9

20.

For x-intercept set $y = 0$

so $0 = 2\log_5(x+3)$

$\Rightarrow \log_5(x+3) = 0$

$\Rightarrow x + 3 = 5^0$

$\Rightarrow x + 3 = 1 \Rightarrow x = -2$

intercept is $(-2,0)$

Choice D

2 marks

- There are several general results used in this question:

1. For x-intercepts set $y = 0$

2. $\log_b a = c \leftrightarrow a = b^c$
   (log statement) (exponential statement)

3. $a^0 = 1$

- The factor 2 'vanishes' since you can divide both sides of the equation by 2.

- An alternative strategy would be to work through the choices substituting the x and y values into the equation to see if it makes sense.

HMRN: p 50–51

21.

$$y = \frac{1}{16}x^4 - \frac{1}{8}x^2 + x \qquad \checkmark$$

$$\Rightarrow \frac{dy}{dx} = \frac{1}{4}x^3 - \frac{1}{4}x + 1 \qquad \checkmark$$

The tangent $y = x + c$ has gradient 1
So set $\qquad \checkmark$

$$\frac{dy}{dx} = 1$$

$$\Rightarrow \frac{1}{4}x^3 - \frac{1}{4}x + 1 = 1$$

$$\Rightarrow \frac{1}{4}x^3 - \frac{1}{4}x = 0 \Rightarrow x^3 - x = 0 \qquad \checkmark$$

$$x(x^2 - 1) = 0 \Rightarrow x(x - 1)(x + 1) = 0$$

$$\Rightarrow x = 0 \text{ or } x = 1 \text{ or } x = -1$$

For $x = 0$:

$$y = \frac{1}{16} \times 0^4 - \frac{1}{8} \times 0^2 + 0 = 0 \qquad \checkmark$$

so $y = x + c$ gives $0 = 0 + c \Rightarrow c = 0$
The tangent is $y = x$ with contact point $(0, 0)$

For $x = -1$:

$$y = \frac{1}{16} \times (-1)^4 - \frac{1}{8} \times (-1)^2 + (-1)$$

$$= -\frac{17}{16}$$

so $y = x + c$ gives

$$-\frac{17}{16} = -1 + c \Rightarrow c = -\frac{1}{16}$$

tangent is $y = x - \frac{1}{16}$,

contact point is $\left(-1, -\frac{17}{16}\right)$ $\qquad \checkmark$

For $x = 1$:

$$y = \frac{1}{16} \times 1^4 - \frac{1}{8} \times 1^2 + 1 = \frac{15}{16}$$

so $y = x + c$ gives

$$\frac{15}{16} = 1 + c \Rightarrow c = -\frac{1}{16}$$

tangent is $y = x - \frac{1}{16}$ (same as for $x = -1$)

contact point is $\left(1, \frac{15}{16}\right)$. $\qquad \checkmark$

Strategy
- Evidence that you know to differentiate will gain you this mark.

Differentiation
- Be careful with the fractions here:

$$4 \times \frac{1}{16} = \frac{4}{16} = \frac{1}{4} \text{ and } 2 \times \frac{1}{8} = \frac{2}{8} = \frac{1}{4}$$

Gradient
- Compare $y = mx + c$ and $y = x + c$. This leads to $m = 1$ for the gradient of the tangent line.

Strategy
- This second strategy mark is given for knowing to set the gradient formula equal to 1.

Solving
- Remove fractions first by multiplying both sides of the equation by 4. Although this is a cubic, terms are missing and so is easily factorised once you realise to remove the common factor x.

Calculation
- This processing mark involves a fair amount of calculation and is gained for clearly stating the possible values of c. These are:

$$c = 0 \text{ and } c = -\frac{1}{16}$$

Interpretation
- With there being three points of contact:

$$(0,0), \left(-1, -\frac{17}{16}\right) \text{ and } \left(1, \frac{15}{16}\right)$$

but only two equations for the tangent:

$$y = x \text{ and } y = x - \frac{1}{16}$$

a close examination of the graph shows $y = x - \frac{1}{16}$ is a tangent at two separate points on the curve as is shown in this diagram

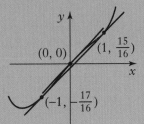

HMRN: p 19–20

7 marks

22. (a)
$f(x) = x^3 + 3x^2 - 4$ ✓
$\Rightarrow f'(x) = 3x^2 + 6x$
For stationary points set
$f'(x) = 0$ ✓
$\Rightarrow 3x^2 + 6x = 0 \Rightarrow 3x(x + 2) = 0$ ✓ ✓
$\Rightarrow x = 0$ or $x = -2$ ✓

$$x: \quad \underset{-2 \qquad 0}{\underline{}}$$

$f'(x) = 3x(x + 2): \quad + \quad - \quad +$

Shape of graph : ╱ ╲ ╱

nature: max   min

$f(-2) = (-2)^3 + 3 \times (-2)^2 - 4 = 0$
So $(-2, 0)$ is a maximum
stationary point
$f(0) = 0^3 + 3 \times 0^2 - 4 = -4$ ✓
So $(0, -4)$ is a minimum
stationary point ✓

6 marks

Differentiate
- Correct differentiation will gain this mark.

Strategy
- To find the stationary points set $f'(x) = 0$.

Solutions
- The common factor is $3x$ with two roots: 0 and -2.

Justify
- A 'nature table' is needed to determine the type (maximum or minimum) of each stationary point.

y-coordinates
- Take care with negatives. In paper 1 questions you have no access to a calculator. In general remember:

  squaring produces positive or zero answers, cubing a negative quantity gives a negative answer. In this case $(-2)^3 = -8$.

Statements
- Clear statements concerning the nature of the points, i.e. maximum or minimum etc. are expected for this final mark.

  HMRN: p 20–21

22. (b) (i) ✓

$$\begin{array}{r|rrrr} -2 & 1 & 3 & 0 & -4 \\ & & -2 & -2 & 4 \\ \hline & 1 & 1 & -2 & 0 \end{array}$$

0 remainder
so $f(-2) = 0$
so $x + 2$ is a
factor ✓

2 marks

Strategy
- You should know to use $x = -2$ to show that $x + 2$ is a factor.

Calculation and conclusion
- Producing a zero in the 'synthetic division scheme' means that dividing by $x + 2$ gives a zero remainder and so shows that $x + 2$ is a factor.

  HMRN: p 25–26

22. (b) (ii)
$x^3 + 3x^2 - 4$
$= (x + 2)(x^2 + x - 2)$ ✓
$= (x + 2)(x + 2)(x - 1)$ ✓
$= (x + 2)^2(x - 1)$ ✓

3 marks

Interpretation
- If $x + 2$ is a factor then the factorisation will give: $(x + 2)$ (other factor).

Quadratic factor
- The row: 1 1 −2 in the 'synthetic division scheme' gives the coefficients of the 'other factor' i.e. $x^2 + x - 2$.

Complete factorisation
- $(x + 2)(x^2 + x - 2)$ is not a *complete* factorisation.

  HMRN: p 25–26

22. (c)

For x-intercepts set $y = 0$

So $x^3 + 3x^2 - 4 = 0$

$\Rightarrow (x - 1)(x + 2)^2 = 0$

$\Rightarrow x = 1$ or $x = -2$

Intercepts are $(-2, 0)$ and $(1, 0)$ ✓

For y-intercept set $x = 0$

So $y = 0^3 + 3 \times 0^2 - 4 = -4$

Intercept is $(0, -4)$ ✓

Sketch:

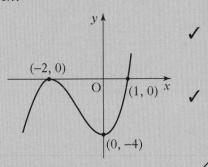

✓

✓

4 marks

x-intercepts
- You should always be aware that previous parts of questions are likely to be used in subsequent parts. In this case (b)(ii) covered the factorisation of $x^3 + 3x^2 - 4$ which is vital for calculating the two x-intercepts $(-2,0)$ and $(1,0)$.

y-intercept
- 1 mark is allocated for this result.

Sketch
- 1 mark is allocated for your sketch showing the cubic shape correctly with the two stationary points clearly shown and labelled.

- The 2nd mark is given for your sketch clearly showing the intercepts, i.e. $(0,-4)$ and $(1,0)$ etc.

HMRN: p 21

23. (a)

$f(x) = 2x - 1$, $g(x) = \log_{12}x$

$f(g(x)) + g(f(x)) = 0$ ✓

$\Rightarrow f(\log_{12}x) + g(2x - 1) = 0$ ✓

$\Rightarrow 2\log_{12}x - 1 + \log_{12}(2x - 1) = 0$ ✓

$\Rightarrow \log_{12}x^2 + \log_{12}(2x - 1) = 1$

$\Rightarrow \log_{12}x^2(2x - 1) = 1$ ✓

$\Rightarrow x^2(2x - 1) = 12^1$ ✓

$\Rightarrow 2x^3 - x^2 - 12 = 0$

$$
\begin{array}{r|rrrr}
2 & 2 & -1 & 0 & -12 \\
  &   & 4 & 6 & 12 \\
\hline
  & 2 & 3 & 6 & 0
\end{array}
$$

So $x = 2$ satisfies the equation and is therefore a solution ✓

6 marks

Composition
- Combining f and g to get the formula $f(g(x))$ is called the 'composition' of f and g. Showing either $f(\log_{12}x)$ or $g(2x - 1)$ gains you this first mark.

Composition
- Reaching either $2\log_{12}x - 1$ or $\log_{12}(2x - 1)$ will gain you this 2nd mark.

Composition
- Correctly finding the 2nd of either $2\log_{12}x - 1$ or $\log_{12}(2x - 1)$ will give you this 3rd mark.

Log Law
- The rule used here is:

$$\log_b m + \log_b n = \log_b mn$$

Exponential form
- The result used is:

$$\log_b a = c \quad \leftrightarrow \quad a = b^c$$
(logarithmic form)   (exponential form)

Calculation
- At this stage you must clearly show that $x = 2$ is a solution of $2x^3 - x^2 - 12 = 0$. The table shows that $f(2) = 0$ where $f(x) = 2x^3 - x^2 - 12$.

HMRN: p 10, p 25, p 50

Strategy
- One solution, $x = 2$, comes from the factor $x - 2$ equating to zero. Any other solutions will therefore come from equating the other factor to zero, i.e. $2x^2 + 3x + 6 = 0$. If your working shows you knew this you will gain this strategy mark.

23. (*b*)
From part (*a*) equation becomes:
$(x - 2)(2x^2 + 3x + 6) = 0$
Consider $2x^2 + 3x + 6 = 0$ ✓
Discriminant $= 3^2 - 4 \times 2 \times 6$
$= 9 - 48 = -39$
Since Discriminant < 0 there are no Real solutions. ✓
The only Real solution is $x = 2$.

2 marks

Communication
- There must be a clear reason for 'no Real solutions'. In this case the discriminant of the quadratic equation is negative. Your working should state this fact quite clearly. The result used is:

Discriminant $< 0 \Rightarrow$ no Real roots.

HMRN: p 26–27

WORKED ANSWERS: EXAM B   **PAPER 2**

1. (*a*) (i)
$x^2 + y^2 + 4x - 6y + 5 = 0$
Centre is $(-2, 3)$.   ✓

1 mark

Centre
• The result used is:

$$x^2 + y^2 + 2gx + 2fy + c = 0$$

    Centre: $(-g, -f)$

This result is on your formulae sheet in the exam.

HMRN: p 39

1. (*a*) (ii)
Radius $= \sqrt{(-2)^2 + 3^2 - 5}$
       $= \sqrt{4 + 9 - 5} = \sqrt{8} = 2\sqrt{2}$ ✓

1 mark

Radius
• Notice that

$$\sqrt{8} = \sqrt{4 \times 2} = \sqrt{4} \times \sqrt{2} = 2 \times \sqrt{2} = 2\sqrt{2}$$

• The radius formula: $\sqrt{g^2 + f^2 - c}$ for the circle $x^2 + y^2 + 2gx + 2fy + c = 0$ is given to you on your formulae sheet in the exam.

HMRN: p 39

1. (*b*) (i)
Circle B:
     $(x - 2)^2 + (y + 1)^2 = 2$
has centre $(2, -1)$ ✓
Distance between $(-2, 3)$ and $(2, -1)$ is:

$\sqrt{(-2 - 2)^2 + (3 - (-1))^2}$
$= \sqrt{(-4)^2 + 4^2} = \sqrt{32} = 4\sqrt{2}$ ✓

2 marks

Centre
• The result used is:

For circle $(x - a)^2 + (y - b)^2 = r^2$ the centre is (a, b) a result also given to you in your exam.

Calculation
• The distance formula is used here. If A (x_1, y_1) and B(x_2, y_2) then

$$AB = \sqrt{(x_2 - x_1)^2 + (y_2 - y_1)^2}$$

HMRN: p 7, p 39

Strategy

- I $\quad$ Distance between centres is greater than the sum of the radii.

II $\quad$ Distance between centres is equal to the sum of the two radii.

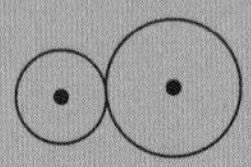

III $\quad$ Distance between centres is less than the sum of the two radii.

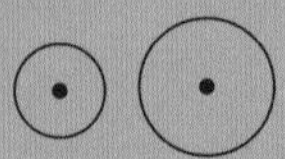

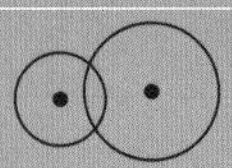

1. (*b*) (ii)

Circle A: radius $= 2\sqrt{2}$

Circle B: radius $= \sqrt{2}$

Distance between centres $= 4\sqrt{2}$

Sum of radii $= 2\sqrt{2} + \sqrt{2} = 3\sqrt{2}$ ✓

less than $4\sqrt{2}$, the distance between centres

$\Rightarrow$ circles do not intersect ✓

2 marks

Communication

- There must be a comparison given. In this case a statement that the sum

$\left(3\sqrt{2}\right)$ is less than the centre distance

$\left(4\sqrt{2}\right)$ so there is no intersection.

This is situation I shown above.

HMRN: p 41

1. (*c*)

Solve $\quad y = x + 5$

$x^2 + y^2 + 4x - 6y + 5 = 0$

Substitute $y = x + 5$ in circle equation: ✓

$x^2 + (x+5)^2 + 4x - 6(x+5) + 5 = 0$ ✓

$\Rightarrow x^2 + x^2 + 10x + 25 + 4x - 6x - 30 + 5 = 0$

$\Rightarrow 2x^2 + 8x = 0 \Rightarrow 2x(x+4) = 0$ ✓

$\Rightarrow x = 0$ or $x = -4$ ✓

when $x = -4$, $y = -4 + 5 = 1$

when $x = 0$, $y = 0 + 5 = 5$

So P(−4, 1) and Q(0, 5) ✓

5 marks

Strategy

- In general to find where graphs $y = f(x)$ and $y = g(x)$ intersect then you equate the formulae i.e. $f(x) = g(x)$ and solve the resulting equation. However in this case the equation of the circle cannot be written as "$y = ...$" so substitution is used.

Substitution

- Replace all occurrences of y by $x + 5$ in the circle equation.

'Standard form'

- Reducing the equation to $2x^2 + 8x = 0$ gains you this mark.

Solve for x

- The common factor is $2x$ with roots −4 and 0.

Coordinates

- Coordinates are asked for not just values of x and y.

HMRN: p 40

2. (a)

A$(-5, 2)$ and B$(-3, 8)$ ✓

$$\Rightarrow m_{AB} = \frac{8-2}{-3-(-5)} = \frac{6}{2} = 3$$

$$\Rightarrow m_\perp = -\frac{1}{3}$$ ✓

Midpoint of AB is

$$\left(\frac{-5+(-3)}{2}, \frac{2+8}{2}\right) = (-4, 5)$$ ✓

For the perpendicular bisector:
A point on the line is $(-4, 5)$
and the gradient $= -\frac{1}{3}$
So equation is

$$y - 5 = -\frac{1}{3}(x-(-4))$$

$$\Rightarrow 3y - 15 = -(x+4)$$
$$\Rightarrow 3y - 15 = -x - 4$$
$$\Rightarrow 3y + x = 11$$ ✓

4 marks

Strategy
- Here you are using the gradient formula:

  P(x_1, y_1), Q (x_2, y_2) gives $m_{PQ} = \frac{y_2 - y_1}{x_2 - x_1}$

Perpendicular Gradient
- If $m = \frac{a}{b}$ then $m_\perp = -\frac{b}{a}$. For $m = 3$ you think of 3 as $\frac{3}{1}$. Inverting and changing sign then gives $-\frac{1}{3}$ as shown in the solution.

Strategy
- You have to know that 'bisector' means the mid point of AB.
- The mid point formula is:

  P(x_1, y_1), Q(x_2, y_2). Midpoint is
  $\left(\frac{x_1 + x_2}{2}, \frac{y_1 + y_2}{2}\right)$

Equation
- Using $y - b = m(x - a)$ with $m = -\frac{1}{3}$ and the point (a, b) is $(-4, 5)$.

  HMRN: p 3–4, p 7

2. (b)

A$(-5, 2)$ and D$(5, -8)$
Midpoint of AD is

$$\left(\frac{-5+5}{2}, \frac{2+(-8)}{2}\right) = M(0, -3)$$ ✓

So using C$(3, 6)$ and M$(0, -3)$

$$m_{CM} = \frac{6-(-3)}{3-0} = \frac{9}{3} = 3$$ ✓

For the median:
A point on the line is $(0, -3)$ and the gradient is 3
So equation is $y - (-3) = 3(x - 0)$
$$\Rightarrow y + 3 = 3x$$
$$\Rightarrow y - 3x = -3$$ ✓

3 marks

Strategy
- The median is the line from C to the midpoint of the opposite side AD. You will therefore need to find the coordinates of the midpoint.

Gradient
- Medians do not normally involve 'perpendicular' and so when m_{CM} is calculated you use this value, 3, to find the equation of the median.

Equation
- Use $y - b = m(x - a)$ with $m = 3$ and (a, b) being the point $(0, -3)$. Alternatively spot that $(0, -3)$ is the y-intercept of the line and use $y = mx + c$ with $m = 3$ and $c = -3$ to give $y = 3x - 3$.

  HMRN: p 7

2. (c)
To find the intersection
    point S: ✓
Solve:

$\left.\begin{array}{l}3y + x = 11 \\ y - 3x = -3\end{array}\right\}_{(\times 3)} \rightarrow \begin{array}{l}3y + x = 11 \\ 3y - 9x = -9\end{array}$

subtract: $10x = 20$
$\Rightarrow x = 2$ ✓

now substitute $x = 2$ in
$y - 3x = -3$
$\Rightarrow y - 3 \times 2 = -3 \Rightarrow y - 6 = -3$
$\Rightarrow y = 3$ ✓
    so S(2,3)

3 marks

Strategy
- To find the point of intersection of two lines you solve the two equations of the lines simultaneously.

Find one variable
- An alternative is to multiply the 1st equation by 3 then add to give $10y = 30 \Rightarrow y = 3$.

Second variable
- Use the 'easier' equation when doing the substitution. If you found $y = 3$ first then the 1st equation is 'easier'.

HMRN: p 6

3.
$3\cos 2x° + 9\cos x° = \cos^2 x° - 7$
$\Rightarrow 3(2\cos^2 x° - 1) + 9\cos x° = \cos^2 x° - 7$ ✓
$\Rightarrow 6\cos^2 x° - 3 + 9\cos x° = \cos^2 x° - 7$
$\Rightarrow 5\cos^2 x° + 9\cos x° + 4 = 0$ ✓
$\Rightarrow (5\cos x° + 4)(\cos x° + 1) = 0$ ✓
$\Rightarrow 5\cos x° + 4 = 0$ or $\cos x° + 1 = 0$
$\Rightarrow \cos x° = -\dfrac{4}{5}$ or $\cos x° = -1$ ✓

For $\cos x° = -\dfrac{4}{5}$:

$x°$ is in 2nd or 3rd quadrants
1st quadrant angle is $36·9°$
so $x = 180 - 36·9$ or
    $x = 180 + 36·9$
$\Rightarrow x = 143·1$ or $x = 216·9$

(continued next page)

Strategy
- There are three expansions of $\cos 2x°$:
$\cos 2x° = 2\cos^2 x° - 1$ or $\cos^2 x° - \sin^2 x°$ or $1 - 2\sin^2 x°$. Which of these you use is dictated by the surrounding 'landscape' in the equation: There is a '$\cos x°$ term' and a $\cos^2 x°$ term but no '$\sin x°$ term' or '$\sin^2 x°$ term'. So $\cos 2x° = 2\cos^2 x° - 1$ is used as the other two forms involve $\sin^2 x°$.

'Standard form'.
- You recognise the equation as a quadratic equation in $\cos x°$ and arrange it into the standard form $5\cos^2 x° + 9\cos x° + 4 = 0$.

Factorisation
- Compare $5c^2 + 9c + 4 = (5c + 4)(c + 1)$.
- Remember that you should always check your answer in a factorisation by multiplying out, i.e. working backwards.

Solving for $\cos x°$
- From $(5c + 4)(c + 1) = 0$ to $c = -\dfrac{4}{5}$ or $c = -1$ is no different to what you do at this stage except that the single variable c is replaced by the expression $\cos x°$.

3. Continued.

For $\cos x° = -1$

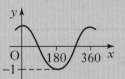

So $x = 180$

Solutions are:
143·1, 180, 216·9
(to 1 decimal place). ✓

5 marks

Solutions
- A lot of knowledge and work needed for this final processing mark!
- The quadrant diagram is used:

$$
\begin{array}{c|c}
\checkmark S & A \\
\hline
\checkmark T & C
\end{array}
$$

for $\cos x°$ negative.

HMRN: p37

4. (a)

$M(0,3,2)$ ✓
$N(5,2,0)$ ✓

2 marks

Point M
- M is the midpoint. Half-way along DG which is 6 units long (y-coordinate) is 3 units.

Point N
- N is $\frac{1}{3}$ of the way along AB so $\frac{1}{3}$ of $6 = 2$ units is the y-coordinate.

HMRN: p 42

4. (b)

$M(0,3,2)$ and $B(5,6,0)$

$$\overrightarrow{MB} = \boldsymbol{b} - \boldsymbol{m}$$

$$= \begin{pmatrix} 5 \\ 6 \\ 0 \end{pmatrix} - \begin{pmatrix} 0 \\ 3 \\ 2 \end{pmatrix} = \begin{pmatrix} 5 \\ 3 \\ -2 \end{pmatrix}$$ ✓

also $M(0,3,2)$ and $N(5,2,0)$

$$\overrightarrow{MN} = \boldsymbol{n} - \boldsymbol{m}$$

$$= \begin{pmatrix} 5 \\ 2 \\ 0 \end{pmatrix} - \begin{pmatrix} 0 \\ 3 \\ 2 \end{pmatrix} = \begin{pmatrix} 5 \\ -1 \\ -2 \end{pmatrix}$$ ✓

2 marks

Components
- The basic result used is:

$P(x_1, y_1, z_1)$ and $Q(x_2, y_2, z_2)$

$$\overrightarrow{PQ} = \boldsymbol{q} - \boldsymbol{p} = \begin{pmatrix} x_2 \\ y_2 \\ z_2 \end{pmatrix} - \begin{pmatrix} x_1 \\ y_1 \\ z_1 \end{pmatrix} = \begin{pmatrix} x_2 - x_1 \\ y_2 - y_1 \\ z_2 - z_1 \end{pmatrix}$$

HMRN: p 43–44

4. (*c*)

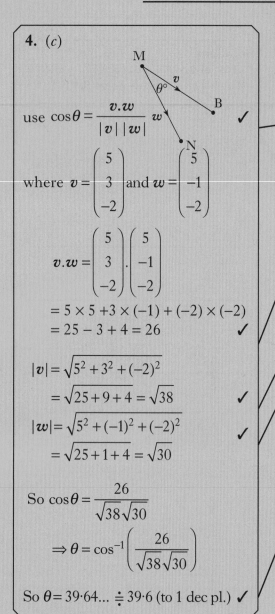

use $\cos\theta = \dfrac{\boldsymbol{v}.\boldsymbol{w}}{|\boldsymbol{v}||\boldsymbol{w}|}$ ✓

where $\boldsymbol{v} = \begin{pmatrix} 5 \\ 3 \\ -2 \end{pmatrix}$ and $\boldsymbol{w} = \begin{pmatrix} 5 \\ -1 \\ -2 \end{pmatrix}$

$\boldsymbol{v}.\boldsymbol{w} = \begin{pmatrix} 5 \\ 3 \\ -2 \end{pmatrix}.\begin{pmatrix} 5 \\ -1 \\ -2 \end{pmatrix}$

$= 5 \times 5 + 3 \times (-1) + (-2) \times (-2)$

$= 25 - 3 + 4 = 26$ ✓

$|\boldsymbol{v}| = \sqrt{5^2 + 3^2 + (-2)^2}$

$= \sqrt{25 + 9 + 4} = \sqrt{38}$ ✓

$|\boldsymbol{w}| = \sqrt{5^2 + (-1)^2 + (-2)^2}$ ✓

$= \sqrt{25 + 1 + 4} = \sqrt{30}$

So $\cos\theta = \dfrac{26}{\sqrt{38}\sqrt{30}}$

$\Rightarrow \theta = \cos^{-1}\left(\dfrac{26}{\sqrt{38}\sqrt{30}}\right)$

So $\theta = 39{\cdot}64... \doteq 39{\cdot}6$ (to 1 dec pl.) ✓

5 marks

Strategy
- This strategy is for the use of the 'scalar' or 'dot' product formula.

Calculation
- The 'dot product' formula is:

$$\begin{pmatrix} x_1 \\ y_1 \\ z_1 \end{pmatrix}.\begin{pmatrix} x_2 \\ y_2 \\ z_2 \end{pmatrix} = x_1x_2 + y_1y_2 + z_1z_2$$

Magnitudes
- The magnitude formula is

$$\left|\left|\begin{pmatrix} x_1 \\ y_1 \\ z_1 \end{pmatrix}\right|\right| = \sqrt{x_1^2 + y_1^2 + z_1^2}$$

Angle
- Be careful with your calculator calculation

$$\boxed{\text{INV}}\,\boxed{\cos}\,\boxed{(}\,\boxed{2}\,\boxed{6}\,\boxed{\div}\,\boxed{(}\,\boxed{\surd}\,\boxed{3}\,\boxed{8}\,\boxed{\times}\,\boxed{\surd}\,\boxed{3}\,\boxed{0}\,\boxed{)}\,\boxed{)}\,\boxed{\text{EXE}}$$

The brackets are vital: $\cos^{-1}(...)$ and $(\sqrt{38} \times \sqrt{30})$.

HMRN: p 46

5. (*a*)

$m = 1$ and $n = \sqrt{3}$ ✓ ✓

2 marks

Interpret graphs
- The 'normal' sine graph has amplitude $= 1$ This gives $m = 1$. The amplitude of the cosine graph shown is $\sqrt{3}$ so $n = \sqrt{3}$.

HMRN: p16–17

5. (b)

$f(x) = \sin x$ and $g(x) = \sqrt{3} \cos x$

so $f(x) - g(x) = \sin x - \sqrt{3} \cos x$

let $\sin x - \sqrt{3} \cos x$

$\qquad = k \sin(x - a), k > 0$

$\Rightarrow \sin x - \sqrt{3} \cos x$

$= k \sin x \cos a - k \cos x \sin a$ ✓

now equate coefficients of $\sin x$ and $\cos x$:

$\left.\begin{array}{l} k \cos a = 1 \\ k \sin a = \sqrt{3} \end{array}\right\}$ since both $\sin a$ and $\cos a$ are positive a is in 1st quadrant ✓

Divide: $\dfrac{k \sin a}{k \cos a} = \dfrac{\sqrt{3}}{1}$

$\Rightarrow \tan a = \sqrt{3}$

$\Rightarrow a = \dfrac{\pi}{3}$ ✓

Square and add:

$(k \cos a)^2 + (k \sin a)^2 = 1^2 + \left(\sqrt{3}\right)^2$

$\Rightarrow k^2 \cos^2 a + k^2 \sin^2 a = 1 + 3$

$\Rightarrow k^2(\cos^2 a + \sin^2 a) = 4$

$\Rightarrow k^2 \times 1 = 4 \Rightarrow k = 2 \ (k > 0)$ ✓

So $f(x) - g(x) = \sin x - \sqrt{3} \cos x$

$\qquad = 2 \sin\left(x - \dfrac{\pi}{3}\right)$

4 marks

Strategy
- You must clearly show the expansion of $k \sin(x - a)$. To do this you use the addition formulae:
$\sin(A \pm B) = \sin A \cos B \pm \cos A \sin B$
which is given on your formulae sheet.

Compare coefficients

compare $\enclose{circle}{1} \sin x \qquad - \enclose{circle}{\sqrt{3}} \cos x$

$\enclose{circle}{k \sin x \cos a} - \enclose{circle}{k \cos x \sin a}$

giving $k \cos a = 1$ and $k \sin a = \sqrt{3}$

Find a
- You should recognise $\sqrt{3}$ as an exact value leading to the angle $\frac{\pi}{3}$.

Find k
- For finding a you used $\frac{\sin a}{\cos a} = \tan a$ and for finding k you use $\sin^2 a + \cos^2 a = 1$. If you try to apply a learnt 'formula' for finding k sometimes mistakes creep in. It is better to understand that squaring and adding both sides of the two equations leads to the value of k and just 'do the mathematics' at the time.

- $k^2 = 4$ giving $k = -2$ is not an allowable value since $k > 0$.

HMRN: p 53–54

5. (c)

$$y = 2\sin\left(x - \frac{\pi}{3}\right)$$

$$\Rightarrow \frac{dy}{dx} = 2\cos\left(x - \frac{\pi}{3}\right) \qquad \checkmark$$

For a gradient of 2 set $\dfrac{dy}{dx} = 2$

$$\Rightarrow 2\cos\left(x - \frac{\pi}{3}\right) = 2$$

$$\Rightarrow \cos\left(x - \frac{\pi}{3}\right) = 1$$

$$\Rightarrow x - \frac{\pi}{3} = 0 \text{ (or } 2\pi)$$

$$\Rightarrow x = \frac{\pi}{3}\left(\text{or } 2\pi + \frac{\pi}{3}\right)$$

But $0 \leq x \leq \pi$ so $x = \dfrac{\pi}{3}$
is the only solution $\qquad \checkmark$

2 marks

Strategy
- You are told the gradient so you have to differentiate and find the x-value that satisfies $\frac{dy}{dx} = 2$. The word 'hence' is very important. It is telling you to use the previous result. You will lose marks if you do not do this.

Differentiation involves the chain rule:

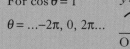
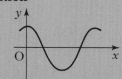

$y = 2\sin(g(x)) \Rightarrow \frac{dy}{dx} = 2\cos(g(x)) \times g'(x)$
In this case $g(x) = x - \frac{\pi}{3}$ so $g'(x) = 1$ so
$\frac{dy}{dx} = 2\cos\left(x - \frac{\pi}{3}\right) \times 1 = 2\cos\left(x - \frac{\pi}{3}\right)$.

Equation and solution
- For $\cos\theta = 1$

  $\theta = \ldots -2\pi,\ 0,\ 2\pi\ldots$

- In this question the angle is $x - \dfrac{\pi}{3}$ so to finally find x you must add $\dfrac{\pi}{3}$.

  HMRN: p 48

6.

Intersection with $y = 15$ solve:

$\left.\begin{array}{l} y = 15 \\ y = x^4 - 1 \end{array}\right\}$ so $x^4 - 1 = 15$
$\Rightarrow x^4 = 16 \Rightarrow x = 2$ or -2

For x-intercept set $y = 0 \qquad \checkmark$
so $x^4 - 1 = 0 \Rightarrow x^4 = 1$
$\qquad \Rightarrow x = 1$ or $-1 \qquad \checkmark$

Here is the diagram:

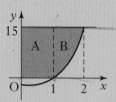

Area A is a rectangle
$\qquad$ Area A $= 1 \times 15 = 15$ unit^2

(continued to next page)

Limits
- It is essential to find the x values of the intersection of the lines $y = 15$ and x-axis with the curve. These values will be used later in the integral for finding the area between two graphs.

x-values
- The relevant values are $x = 1$ and $x = 2$. Notice that all the work can be done in the 1st quadrant and then doubled at the end since the y-axis is an axis of symmetry for the diagram.

6. Continued

Area B =

$$= \int_1^2 15 - (x^4 - 1) \, dx \quad ✓$$

$$= \int_1^2 16 - x^4 dx = \left[16x - \frac{x^5}{5} \right]_1^2 \quad ✓$$

$$= \left(16 \times 2 - \frac{2^5}{5} \right) - \left(16 \times 1 - \frac{1^5}{5} \right) \quad ✓$$

$$= 32 - \frac{32}{5} - 16 + \frac{1}{5} = 16 - \frac{31}{5} = \frac{49}{5} \text{unit}^2$$

$$✓$$

Area A + Area B

$$= 15 + \frac{49}{5} = \frac{75}{5} + \frac{49}{5} = \frac{124}{5} \text{unit}^2 \quad ✓$$

By symmetry the required area

$$= 2 \times \frac{124}{5} = \frac{248}{5} \text{unit}^2 \quad ✓$$

8 marks

Strategy
- You should know to use integration to find the area between $y = 15$ and the curve (area B) and add the area of Rectangle A. There are other methods possible.

Integration
- Notice
$$\int a \, dx = ax + c \text{ and } \int x^n = \frac{x^{n+1}}{n+1} + c$$
(a is a constant)
- The constant c is not needed when there are limits.

Limits
- A mark is allocated for correct use of limits 1 and 2.

Evaluate
- Using $\int_a^b f(x) dx = [F(x)]_a^b = F(b) - F(a)$
Where $F(x)$ is the result of integrating $f(x)$.

Strategy
- Knowing what to add together!

Calculation
- Final answer is gained by doubling as you have only found the area in the 1st quadrant.

HMRN: p 32–33

---

7. (a)

(0,9) lies on the curve
so $y = 9$ when $x = 0$
$y = -x^2 + a$ gives $9 = -0^2 + a$
$\Rightarrow a = 9$. ✓

1 mark

Calculation
- If (p,q) lies on a graph with equation $y = f(x)$ then $q = f(p)$. In other words the coordinates of the point can be substituted into the equation. In this case this gives the value of a.

HMRN: p 9

---

7. (b)

$f(x) = -x^2 + 9 = 9 - x^2$
to find coordinates of P set $x = m$
so $f(m) = 9 - m^2 \Rightarrow P(m, 9 - m^2)$
$AP = 9 - m^2$ ✓

1 mark

Calculation of y-coordinate
- As can be seen from this diagram, the length AP is the y-coordinate of P. The x-coordinate of P you know is m, i.e. $x = m$.

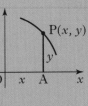

HMRN: p 9

7. (*c*)

The area of the rectangle, $A(m)$ is given by:

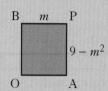

B $\quad m \quad$ P

$9 - m^2$

O $\qquad$ A

$A(m) = m\,(9 - m^2)$ ✓

$= 9m - m^3 \Rightarrow A'(m) = 9 - 3m^2$ ✓

For stationary value set $A'(m) = 0$

✓

$\Rightarrow 9 - 3m^2 = 0 \Rightarrow 3m^2 = 9$

$\Rightarrow m^2 = 3$

So $m = \sqrt{3}$, $m \neq -\sqrt{3}$ since $m > 0$ ✓

$x:$ $\qquad \sqrt{3} \qquad$

$A'(m) = 9 - 3m^2:$ $\quad + \quad - $

Shape of graph: ╱ ‾ ╲ ✓

nature: $\qquad$ max

So $m = \sqrt{3}$ gives a maximum value for the area of rectangle OAPB. ✓

6 marks

Communication
- Area of a rectangle is given by length × breadth and in this case the dimensions are m units × $(9 - m^2)$ units.

Differentiate
- The function $A(m)$ gives the area of the rectangle whereas $A'(m)$ is the gradient function and determines the slope of the graph $y = A(m)$ showing the various areas as m changes.

Strategy
- Setting $A'(m)$ to zero determines which values of m give stationary points on the Area graph.

Solve
- Here there is a positive and a negative value of m. Only the positive value makes sense as you are told that $0 \leq m \leq 3$.

Justify
- The 'nature' table is required to show that $m = \sqrt{3}$ does give a maximum value for the Area.

Communication
- A clear statement summarising your findings is needed for this final mark.

HMRN: p 21–22

---

7. (*d*)

$A\left(\sqrt{3}\right) = \sqrt{3}\left(9 - \left(\sqrt{3}\right)^2\right)$

$= \sqrt{3}(9 - 3)$

$= \sqrt{3} \times 6$ ✓

So $6\sqrt{3}$ unit^2 is the maximum area.

1 mark

Calculation
- When you read the word 'exact' you know to steer clear of decimal approximations. Your answer will be an integer value, a fraction (rational number) or a surd (involving roots) or perhaps an expression involving π or e, but never approximate decimals.

HMRN: p 20

WORKED ANSWERS: EXAM C **PAPER 1**

1.
Let the limit be L
then $L = 0.9L - 1$
$\Rightarrow L - 0.9L = -1 \Rightarrow 0.1L = -1$
So $L = \dfrac{-1 \ (\times 10)}{0.1 \ (\times 10)} = \dfrac{-10}{1} = -10$

Choice A

2 marks

- Remember that limits only exist if the multiplier, in this case 0.9, lies between -1 and 1.
- No calculator is available during Paper 1 so you have to be able to calculate $-\dfrac{1}{0.1}$. The easiest was is to multiply 'top' and 'bottom' by 10.

HMRN: p24

2.
A$(2k, 3)$ and B$(k, 5)$
$\Rightarrow m_{AB} = \dfrac{3-5}{2k-k} = \dfrac{-2}{k} = -\dfrac{2}{k}$

but $m_{AB} = 4$ So $-\dfrac{2}{k} = 4$

$\Rightarrow -2 = 4k \Rightarrow k = -\dfrac{2}{4} = -\dfrac{1}{2}$

Choice B

2 marks

- The gradient formula is:
  If $P(x_1, y_1)$, $Q(x_2, y_2)$ then $m_{PQ} = \dfrac{y_2 - y_1}{x_2 - x_1}$.
  You are not given this formula during the exam.
- $-\dfrac{2}{k} = 4$ multiply both sides by k
  so $-\dfrac{2}{k_1} \times k^1 = 4 \times k \Rightarrow -2 = 4k$ etc....
- It is possible to work through the choices, in each case finding points A and B and calculating the gradient. This will be more time consuming.

HMRN: p3

3.
$x^2 - 4 = (x-2)(x+2)$ so both
$(x-2)$ and $(x+2)$ are factors
Since $f(-1) = 0$, $(x+1)$ is a factor
so $f(x) = (x+1)(x-2)(x+2)$

Choice D

2 marks

- The 'difference of squares' pattern is:
  $$A^2 - B^2 = (A-B)(A+B)$$
  In this case you have $x^2 - 2^2$
- For a polynomial equation $f(x) = 0$ there is a correspondence between roots and factors:
  -1 is a root $\leftrightarrow x + 1$ is a factor
  $f(-1) = 0$ tells you -1 is a root.

HMRN: p 26

4.
$u_{n+1} = -\dfrac{1}{2} u_{n+1}, u_0 = 4$

So $u_1 = -\dfrac{1}{2} \times 4 + 1 = -2 + 1 = -1$

and $u_2 = -\dfrac{1}{2} \times (-1) + 1$

$= \dfrac{1}{2} + 1 = \dfrac{3}{2}$

Choice C

2 marks

- The process is:
  $u_0 \rightsquigarrow u_1 \rightsquigarrow u_2$
  multiply by $-\frac{1}{2}$ multiply by $-\frac{1}{2}$
  then add 1   then add 1

In this case:
  $4 \rightsquigarrow -1 \rightsquigarrow \frac{3}{2}$
  $4 \times (-\frac{1}{2}) + 1$ $-1 \times (-\frac{1}{2}) + 1$

HMRN: p 23

5.

$2\cos x - \sqrt{2} = 0 \Rightarrow 2\cos x = \sqrt{2}$

$\Rightarrow \cos x = \dfrac{\sqrt{2}}{2} = \dfrac{\sqrt{2}}{\sqrt{2} \times \sqrt{2}} = \dfrac{1}{\sqrt{2}}$

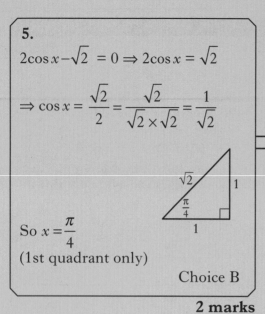

So $x = \dfrac{\pi}{4}$

(1st quadrant only)

Choice B

2 marks

- The initial strategy is to change the equation to the format: $\cos(\text{angle}) = \text{number}$

  in this case $\cos x = \dfrac{\sqrt{2}}{2}$

- You have to recognise an exact value however $\dfrac{\sqrt{2}}{2}$ is not the normal form.

- $\dfrac{\sqrt{2}}{2} = \dfrac{\sqrt{2} \times \sqrt{2}}{2 \times \sqrt{2}} = \dfrac{\cancel{2}^{1}}{_{1}\cancel{2}\sqrt{2}} = \dfrac{1}{\sqrt{2}}$ is an alternative.

  HMRN: p 15, p 34

6.

$4y = -3x + 2$

$\Rightarrow y = -\dfrac{3}{4}x + \dfrac{2}{4}$

So $m = -\dfrac{3}{4} \Rightarrow m_{\perp} = \dfrac{4}{3}$

Choice C

2 marks

- The initial strategy here is to change the equation to the format: $y = mx + c$

  in this case $m = -\dfrac{3}{4}$.

- In general if $m = \dfrac{a}{b}$ then $m_{\perp} = -\dfrac{b}{a}$, the process is to 'invert' the fraction and change the sign. In this case negative changes to positive.

  HMRN: p 4–5

7.

E(1,–1,–1), F(–1,–1,0), G(–7,–1,3)

$\overrightarrow{EF} = f - e = \begin{pmatrix} -1 \\ -1 \\ 0 \end{pmatrix} - \begin{pmatrix} 1 \\ -1 \\ -1 \end{pmatrix} = \begin{pmatrix} -2 \\ 0 \\ 1 \end{pmatrix}$

$\overrightarrow{FG} = g - f = \begin{pmatrix} -7 \\ -1 \\ 3 \end{pmatrix} - \begin{pmatrix} -1 \\ -1 \\ 0 \end{pmatrix} = \begin{pmatrix} -6 \\ 0 \\ 3 \end{pmatrix}$

So $\overrightarrow{FG} = 3\overrightarrow{EF}$

E   F   G

1 part   3 parts

F divides EG in the ratio 1:3

Choice C

2 marks

- If you know three points are collinear then select two different 'journeys' between the points. In the solution $\overrightarrow{EF}$ and $\overrightarrow{FG}$ were selected. The 'position vector' result, e.g. $\overrightarrow{AB} = b - a$ is used to calculate the components of these 'journeys'. Collinearity ensures that one 'journey' is a multiple of the other 'journey'. In this case $\overrightarrow{FG} = 3\overrightarrow{EF}$. Other selections, in this case, would give:

  $\overrightarrow{EF} = \tfrac{1}{4}\overrightarrow{EG}, \overrightarrow{FG} = \tfrac{3}{4}\overrightarrow{EG}, \overrightarrow{GE} = 4\overrightarrow{FE}$ etc.

  Then use this information to sketch the relationship between the points (see diagram in solution) and therefore determine the required ratio.

  HMRN: p 44

8.

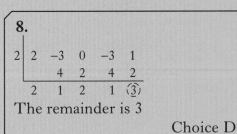

The remainder is 3

Choice D

2 marks

- The 'synthetic division scheme' is used.
- Take care to look for 'missing terms'. In this case there is no 'x^2 term' and that is recorded as a 0 in the top row.
- a | Polynomial coefficients — This means if you divide the polynomial by $x-a$ the remainder is b.

  HMRN: p 25–26

9.

$$(x + 4)(x - 2)$$
$$= x^2 + 2x - 8$$
$$= (x + 1)(x + 1) - 1 - 8$$
$$= (x + 1)^2 - 9$$

Compare: $(x + a)^2 + b$

so $b = -9$

Choice B

2 marks

- The question is trying to confuse you by not giving the quadratic in 'standard form', i.e. with no brackets: $x^2 + 2x - 8$.
- Remember to take your answer, expand it and check it works. In this case:

  $(x + 1)^2 - 9 = x^2 + 2x + 1 - 9 = x^2 + 2x - 8$.
- b is not 9, which is a common mistake. Fortunately that is not one of the choices.

  HMRN: p13

10.

$$f(x) = (1 - x^3)^{1/3}$$
$$\Rightarrow f'(x) = \frac{1}{3}(1 - x^3)^{-2/3} \times (-3x^2)$$
$$= -x^2(1 - x^3)^{-2/3}$$

Choice A

2 marks

- The function given is of the format $f(x) = (g(x))^{1/3}$ where $g(x) = 1 - x^3$. The 'chain rule' is used in this case, so:

  $$f'(x) = \frac{1}{3}(g(x))^{-2/3} \times g'(x)$$

  where $g'(x) = -3x^2$
- Note that

  $$\frac{1}{3} \times (-3x^2) = -\frac{1}{3} \times 3 \times x^2 = -1 \times x^2$$

  HMRN: p 48–49

11.

$$y = -(f(x) + 1) = -f(x) - 1$$
$$y = f(x) \rightarrow y = -f(x)$$

This is reflection in the x-axis

$$y = -f(x) \rightarrow y = -f(x) - 1$$

followed by a translation of 1 unit in the negative direction parallel to the y-axis.

Choice B

2 marks

- An alternative method would be to first consider $y = f(x) + 1$. This is the graph $y = f(x)$ shifted up 1 unit parallel to the y-axis. Then $y = -(f(x) + 1)$ is the new graph 'flipped' in the x-axis.
- Remember to follow your steps using each of the points $(0,0)$ and $(1,-1)$ to check your final choice works for the known points on the graph.

  HMRN: p 9–10

12.

$$m = \frac{1}{2}(a+b) = \frac{1}{2}\left[\begin{pmatrix} -1 \\ 2 \\ 0 \end{pmatrix} + \begin{pmatrix} -2 \\ 3 \\ 1 \end{pmatrix}\right]$$

$$= \frac{1}{2}\begin{pmatrix} -3 \\ 5 \\ 1 \end{pmatrix} = \begin{pmatrix} -3/2 \\ 5/2 \\ 1/2 \end{pmatrix}$$

Choice B

2 marks

- The connection between the coordinates of a point P and its position vector p is as follows:

  $P(a,b,c)$ and $p = \begin{pmatrix} a \\ b \\ c \end{pmatrix}$

  The mid point result for points:

  $P(x_1,y_1,z_1)$, $Q(x_2,y_2,z_2)$

  Mid point $\left(\frac{x_1+x_2}{2}, \frac{y_1+y_2}{2}, \frac{z_1+z_2}{2}\right)$

  works for position vectors

  HMRN: p 44–45

13.

For stationary points set $f'(x) = 0$
$\Rightarrow x^2 + 1 = 0 \Rightarrow x^2 = -1$
and since $x^2 \geq 0$ this equation
has no Real solutions. f has no
stationary points.

Choice A

2 marks

- In this case notice $f'(x)$ is a quadratic. So $f'(x) = 0$ is therefore a quadratic equation. Any solution (root) of this equation gives a stationary point.

- Alternative method: find discriminant of $x^2+1 = 0$. In this case compare $ax^2 + bx + c = 0$ giving $a = 1$, $b = 0$, $c = 1$. Discriminant $= 0^2 - 4 \times 1 \times 1 = -4$. A negative discriminant $\Rightarrow$ no Real roots $\Rightarrow$ no stationary points.

  HMRN: p 20, p 27

14.

$m_{OA} = \frac{3}{2}$ so $\tan a^\circ = \frac{3}{2}$

$\Rightarrow a^\circ = \tan^{-1}\left(\frac{3}{2}\right)$

Choice D

2 marks

- The result used is:

  gradient = $\tan \theta^\circ$

- Alternatively, from the given diagram

  you get $\tan a^\circ = \frac{3}{2}$

- Always use $\tan^{-1}$ to find the angle.

  HMRN: p 4

15.

$\log_4(x^2-4) - 2\log_4(x-2)$
$= \log_4(x^2-4) - \log_4(x-2)^2$

$= \log_4 \frac{x^2-4}{(x-2)^2} = \log_4 \frac{(x-2)(x+2)}{(x-2)(x-2)}$

$= \log_4 \frac{x+2}{x-2}$

Choice C

2 marks

- The 'laws of logs' used are:

  $m \log_b a = \log_b a^m$

  and $\log_b m - \log_b n = \log_b \frac{m}{n}$

- In fractions with algebraic expressions you should always factorise 'top & bottom' expression to see if cancellation can take place. In this case $x - 2$ can be cancelled.

  HMRN: p 50

16.

$\int (2 - 3x)^{\frac{1}{3}} \, dx$

$= \dfrac{(2 - 3x)^{\frac{4}{3}}}{\frac{4}{3} \times (-3)} + c = \dfrac{(2 - 3x)^{\frac{4}{3}}}{-4} + c$

$= -\dfrac{1}{4}(2 - 3x)^{\frac{4}{3}} + c$

Choice A

2 marks

- The integration result used here is called a 'special integral'. The formula used in this case is:

$$\int (ax + b)^n \, dx = \dfrac{(ax+b)^{n+1}}{a(n+1)} + c$$

where $a = -3$, $b = 2$ and $n = \frac{1}{3}$. Notice the division by 'a' the coefficient of x, -3 in this case.

- Note: $\frac{1}{3} + 1 = \frac{1}{3} + \frac{3}{3} = \frac{4}{3}$

  and $\frac{4}{3} \times (-3) = -\dfrac{4 \times \cancel{3}^{1}}{\cancel{3}_{1}} = -4$

- You could, of course, differentiate each choice using the 'chain rule' to determine the correct choice. The 'inverse' of integration is differentiation.

HMRN: p 49

17.

P($-1, 2, 5$) and Q($-3, -1, 4$)

$\overrightarrow{QR} = -2\overrightarrow{PQ} \Rightarrow r - q = -2(q - p)$

$\Rightarrow r - q = -2q + 2p \Rightarrow r = -q + 2p$

So

$r = -\begin{pmatrix} -3 \\ -1 \\ 4 \end{pmatrix} + 2\begin{pmatrix} -1 \\ 2 \\ 5 \end{pmatrix} = \begin{pmatrix} 3 + 2 \times (-1) \\ 1 + 2 \times 2 \\ -4 + 2 \times 5 \end{pmatrix}$

$\Rightarrow r = \begin{pmatrix} 1 \\ 5 \\ 6 \end{pmatrix}$ So R(1, 5, 6)

Choice A

2 marks

- Rewriting a 'journey' like $\overrightarrow{AB}$ in terms of the position vectors a and b of the points A and B allows you to use your algebra skills to solve a question like this. The basic result is:

$$\overrightarrow{AB} = b - a$$

So, in this case, $\overrightarrow{QR} = -2\overrightarrow{PQ}$ is translated into $r - q = -2(q - p)$. The aim is then to solve this equation for r since both p and q are known: $r = -q + 2p$

- If A (x_1, y_1, z_1) then $a = \begin{pmatrix} x_1 \\ y_1 \\ z_1 \end{pmatrix}$

HMRN: p 44–45

18.

$f(x) = -2 \sin 3x$

$f'(x) = -2 \cos 3x \times 3 = -6 \cos 3x$

Choice C

2 marks

- This question involves use of the 'chain rule'. In this case the result used is:

$f(x) = b \sin(g(x)) \Rightarrow f'(x) = b \cos(g(x)) \times g'(x)$ with $b = -2$ and $g(x) = 3x$ so $g'(x) = 3$.

- Your formulae sheet gives:

$f(x)$	$f'(x)$
$\sin ax$	$a \cos ax$

HMRN: p 48–49

19.

$y = k(x - 1)(x + 2)$

$\left(-\frac{1}{2}, 9\right)$ lies on the curve

So $y = 9$ when $x = -\frac{1}{2}$

$\Rightarrow 9 = k\left(-\frac{1}{2} - 1\right)\left(-\frac{1}{2} + 2\right)$

$\Rightarrow 9 = k \times \left(-\frac{3}{2}\right) \times \left(\frac{3}{2}\right)$

$\Rightarrow 9 = -\frac{9}{4}k \Rightarrow 36 = -9k \Rightarrow k = -4$

Choice B

2 marks

- To understand the role k plays in this question here is a diagram showing the graph $y = k(x - 1)(x + 2)$ for different values of k:

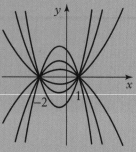

All these curves pass through $(-2, 0)$ and $(1, 0)$, These intercepts corresponding to factors $(x + 2)$ and $(x - 1)$.

Only one curve passes through $\left(-\frac{1}{2}, 9\right)$ and there will be one value of k that gives this curve. Substituting $x = -\frac{1}{2}$ and $y = 9$ in the equation $y = k(x - 1)(x + 2)$ will determine what that value is.

HMRN: p 29

20.

$$\int_0^2 \frac{1}{(4x + 1)^{3/2}} \, dx = \left[-\frac{1}{2(4x + 1)^{1/2}} \right]_0^2$$

$$= -\frac{1}{2\sqrt{4 \times 2 + 1}} - \left(-\frac{1}{2\sqrt{4 \times 0 + 1}} \right)$$

$$= -\frac{1}{2\sqrt{9}} + \frac{1}{2\sqrt{1}} = -\frac{1}{6} + \frac{1}{2}$$

$$= -\frac{1}{6} + \frac{3}{6} = \frac{2}{6} = \frac{1}{3}$$

Choice D

2 marks

- When limits appear on the integral sign the 'constant of integration' c is not required.

- You are using:

$$\int_a^b f(x)\,dx = \left[F(x) \right]_a^b = F(b) - F(a)$$

where $F(x)$ is the result of integrating $f(x)$.

- You should know: $a^{1/2} = \sqrt{a}$

HMRN: p 49

21.

$\sin 2x - \sqrt{3} \sin x = 0$ ✓

$\Rightarrow 2 \sin x \cos x - \sqrt{3} \sin x = 0$

$\Rightarrow \sin x (2\cos x - \sqrt{3}) = 0$ ✓

$\Rightarrow \sin x = 0 \text{ or } \cos x = \dfrac{\sqrt{3}}{2}$ ✓

For $\sin x = 0$
$x = 0, \pi, 2\pi$

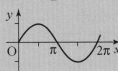

For $\cos x = \dfrac{\sqrt{3}}{2}$ x is in
1st or 4th quadrants

1st quadrant angle is $\dfrac{\pi}{6}$.

So
$x = \dfrac{\pi}{6}$ or $2\pi - \dfrac{\pi}{6} = \dfrac{12\pi}{6} - \dfrac{\pi}{6} = \dfrac{11\pi}{6}$ ✓

Solutions are:

$$x = 0, \frac{\pi}{6}, \pi, \frac{11\pi}{6}, 2\pi.$$

5 marks

Strategy
- Using the 'Double Angle Formula' allows you to factorise the expression. The formula $\sin 2A = 2 \sin A \cos A$ is given to you in the exam.

Factorise
- The common factor is $\sin x$.

Solve
- Notice $2\cos x - \sqrt{3} = 0$

$$\Rightarrow 2\cos x = \sqrt{3} \Rightarrow \cos x = \tfrac{\sqrt{3}}{2}$$

Angles
- For $\sin x$ or $\cos x$ equal to values -1, 0 or 1 you should use the $y = \sin x$ or $y = \cos x$ graph to determine the angles.

- Take care that the angles you give as solutions are allowed. In this case the interval in which the angles lie is $0 \leq x \leq 2\pi$.

- For quadrant use:
  ✓ shows cosine positive

S	A✓
T	C✓

HMRN: p 15, p 37

Strategy
- Evidence that you knew to differentiate will gain you this strategy mark.

Differentiate
- The rule used here is:

$$f(x) = ax^n \Rightarrow f'(x) = nax^{n-1}$$

where a is a constant.

22. (a)

$y = x^3 - 3x^2 - 24x - 28$ ✓

$\Rightarrow \frac{dy}{dx} = 3x^2 - 6x - 24$ ✓

For stationary points set $\frac{dy}{dx} = 0$ ✓

$\Rightarrow 3x^2 - 6x - 24 = 0$

$\Rightarrow 3(x^2 - 2x - 8) = 0$

$\Rightarrow 3(x + 2)(x - 4) = 0$

$\Rightarrow x + 2 = 0$ or $x - 4 = 0$

$\Rightarrow x = -2$ or $x = 4$. ✓

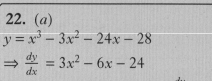

Shape of graph:

nature: max min ✓

When $x = -2$

$y = (-2)^3 - 3 \times (-2)^2 - 24 \times (-2) - 28$

$= -8 - 12 + 48 - 28 = 0$

so $(-2, 0)$ is a maximum stationary point.

When $x = 4$

$y = 4^3 - 3 \times 4^2 - 24 \times 4 - 28$

$= 64 - 48 - 96 - 28 = -108$ ✓

so $(4, -108)$ is a minimum stationary point. ✓

7 marks

Strategy
- The 2nd strategy mark is for setting $\frac{dy}{dx}$ equal to zero. The places on a curve where the gradient is zero are the stationary points.

x-values
- Remember when solving quadratic equations to check for common factors, in this case 3. This reduces the size of the coefficients and makes the subsequent factorisation easier.

Justify
- Evidence has to be given about the 'nature' of the stationary points. There are three types:

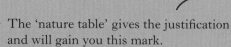

maximum   minimum   Stationary point of inflection

- The 'nature table' gives the justification and will gain you this mark.

y-values.
- There is a mark allocated for the correct calculation of the two y-values for the stationary points.

- Remember to use the 'original' formula $y = \dots$ and not the $\frac{dy}{dx} \dots$ gradient formula for this calculation.

Communication
- Statements must be made stating 'maximum' or 'minimum'.

HMRN: p 20–21

22. (b)

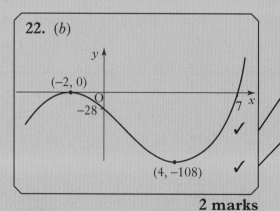

2 marks

Sketch
- 1 mark here will be for showing the correctly shaped graph with the maximum and minimum.

- 1 mark is for the 'annotation', i.e. labelling the points with $(-2, 0)$ and $(4, -108)$.

- Set $x = 0$ in $y = x^3 - 3x^2 - 24x - 28$ for y-intercept.

HMRN: p 21

23. *(a)*

From the diagram

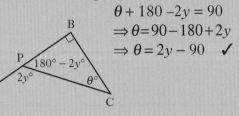

$$\theta + 180 - 2y = 90$$
$$\Rightarrow \theta = 90 - 180 + 2y$$
$$\Rightarrow \theta = 2y - 90 \quad \checkmark$$

Angle

- "Find θ in terms of y." means you are aiming for : $\theta = $ (an expression involving y only) So θ cannot appear on the 'right-hand side'.

- You are using: the angle sum in a triangle is $180°$.

1 mark

Expansion

- The formula:
$\sin(A \pm B) = \sin A \cos B \pm \cos A \sin B$ is given to you in your exam.

23. *(b)*

$$\sin \theta° = \sin(2y - 90)°$$
$$= \sin 2y° \cos 90° - \cos 2y° \sin 90° \quad \checkmark$$
$$= \sin 2y° \times 0 - \cos 2y° \times 1 \quad \checkmark$$
$$= -\cos 2y° = -(2\cos^2 y° - 1) \quad \checkmark$$
$$= -2\cos^2 y° + 1 = 1 - 2\cos^2 y°$$

now $\cos y° = \dfrac{1}{\sqrt{5}}$ $\quad \checkmark$

So $\sin \theta° = 1 - 2 \times \left(\dfrac{1}{\sqrt{5}}\right)^2$ $\quad \checkmark$

$$= 1 - 2 \times \frac{1}{5} = 1 - \frac{2}{5} = \frac{3}{5} \quad \checkmark$$

6 marks

Simplify

- The values $\cos 90° = 0$ and $\sin 90° = 1$ are known from the graphs.

$\cos 90° = 0$

$\sin 90° = 1$

Remember, no calculators in Paper 1.

Strategy

- Use of the 'Double angle' formula.

Calculation

- The exact value of $\cos y°$ is obtained using Pythagoras' Theorem in $\triangle$ APD and then using SOHCAHTOA.

Value

- Substitution of $\dfrac{1}{\sqrt{5}}$ for $\cos y°$ in the expression.

Simplify

- Notice : $\left(\dfrac{1}{\sqrt{5}}\right)^2 = \dfrac{1}{\sqrt{5}} \times \dfrac{1}{\sqrt{5}} = \dfrac{1 \times 1}{\sqrt{5} \times \sqrt{5}} = \dfrac{1}{5}$

HMRN: p 15, p 35–36

24. $\log_{\sqrt{2}} x - \log_{\sqrt{2}} 2 = 2$

$$\Rightarrow \log_{\sqrt{2}} \frac{x}{2} = 2 \quad \checkmark\checkmark$$

$$\Rightarrow \frac{x}{2} = \left(\sqrt{2}\right)^2 = 2 \quad \checkmark$$

$$\Rightarrow x = 4. \quad \checkmark$$

4 marks

Log Law

- The 'law' used is $\log_b m - \log_b n = \log_b \dfrac{m}{n}$.

Exponential form

- Rewrite $\log_b a = c$ as $a = b^c$.

Start Solution

- $\left(\sqrt{2}\right)^2 = \sqrt{2} \times \sqrt{2} = 2$

Finish Solutions

- $\dfrac{x}{2} = 2 \Rightarrow \dfrac{x}{2} \times 2 = 2 \times 2 \Rightarrow x = 4.$

HMRN: p 50

25.

$f'(x) = x^2(x-1) = x^3 - x^2$ ✓

$\Rightarrow f(x) = \int x^3 - x^2 \ dx$ ✓ ✓

$\quad = \dfrac{x^4}{4} - \dfrac{x^3}{3} + c$

now $f(2) = \dfrac{1}{3}$ ✓

$\Rightarrow \dfrac{2^4}{4} - \dfrac{2^3}{3} + c = \dfrac{1}{3}$

$\Rightarrow 4 - \dfrac{8}{3} + c = \dfrac{1}{3}$ ✓

$\Rightarrow c = \dfrac{1}{3} + \dfrac{8}{3} - 4 = 3 - 4 = -1$

so $f(x) = \dfrac{1}{4}x^4 - \dfrac{1}{3}x^3 - 1$

5 marks

Strategy

- In questions where you are given $f'(x)$ (or $\dfrac{dy}{dx}$) and asked to find $f(x)$ (or y) then you need to use integration:

$$f(x) \xrightarrow{\text{differentiation}} f'(x)$$
$$f(x) \xleftarrow{\text{integration}} f'(x)$$

Preparation

- Multiplying out brackets is essential before you integrate.

Integrate

- You are using $\int x^n dx = \dfrac{x^{n+1}}{n+1} + c$.

Substitution

- The 'constant of integration' c is crucial in this question. $f(2) = \frac{1}{3}$ means when $x = 2$ the formula gives $\frac{1}{3}$. c can therefore be found.

Calculation

- No calculator — so practice your fraction work!

HMRN: p31

WORKED ANSWERS: EXAM C **PAPER 2**

1. (a)
$3\sin x° - \cos x° = k\sin(x-a)°$
$\Rightarrow 3\sin x° - \cos x°$
$= k\sin x°\cos a° - k\cos x°\sin a°$ ✓
now equate the coefficients of
$\sin x°$ and $\cos x°$

$\left.\begin{array}{l} k\cos a° = 3 \\ k\sin a° = 1 \end{array}\right\}$ since both $\sin a°$ and $\cos a°$ are positive, $a°$ is in the 1st quadrant. ✓

Divide:

$\dfrac{k\sin a°}{k\cos a°} = \dfrac{1}{3} \Rightarrow \tan a° = \dfrac{1}{3}$

so $a° = \tan^{-1}\left(\dfrac{1}{3}\right) \doteq 18·4°$ ✓

(to 1 decimal place)

Square and add:
$(k\cos a°)^2 + (k\sin a°)^2 = 3^2 + 1^2$
$\Rightarrow k^2\cos^2 a° + k^2\sin^2 a° = 9 + 1$
$\Rightarrow k^2(\cos^2 a° + \sin^2 a°) = 10$
$\Rightarrow k^2 \times 1 = 10 \Rightarrow k^2 = 10$
$\Rightarrow k = \sqrt{10}\ (k>0)$ ✓
So $3\sin x° - \cos x°$
$= \sqrt{10}\sin(x - 18·4)°$

(correct to 1 decimal place)

4 marks

Strategy
- Your first step is to expand $k\sin(x-a)°$.
- On your formulae sheet is:
  $\sin(A±B) = \sin A\cos B ± \cos A\sin B.$
  This is the expansion you should use.
- Notice that there is an implied step in the working:
  $k\sin(x-a)° = k(\sin x°\cos a° - \cos x°\sin a°)$
  $= k\sin x°\cos a° - k\cos x°\sin a°$
- If this expansion does not appear in your working you will not be awarded this mark.

Coefficients
- The method is:
  $\boxed{3}\sin x° \quad -\boxed{1}\cos x°$
  $k\sin x°\cos a° - k\cos x°\sin a°$
  so $k\cos a° = 3$ and $k\sin a° = 1$.

Angle
- A common mistake is to divide in the wrong order. The result you are using is $\dfrac{\sin a°}{\cos a°} = \tan a°$
  In this case sin is on the 'top' of the fraction but comes from the 'bottom' equation so the fraction is $\dfrac{1}{3}$ not $\dfrac{3}{1}$.

Amplitude
- You should try to understand the method for calculating k. This will help you deal with more unusual questions that sometimes arise – these require understanding, not the blind application of a formula.

HMRN: p 53–54

1. (b)

$3\sin x° - \cos x° = 1$

$\Rightarrow \sqrt{10}\ \sin(x - 18·4...)° = 1$ ✓

$\Rightarrow \sin(x - 18·4...)° = \dfrac{1}{\sqrt{10}}$

$\Rightarrow x° - 18·4...° = \sin^{-1}\left(\dfrac{1}{\sqrt{10}}\right)$

$\qquad = 18·4...°$ ✓

so $x = 18·43... + 18·43...$

$\qquad = 36·86...$

$\Rightarrow x \doteq 36·9$ (to decimal place)

$\qquad (0 \le x \le 90)$ ✓

3 marks

Strategy
- Using your result from part (a) you can rewrite the given equation in a form that is easier to solve.

Solve for (x–a)°
- Your aim is to move to an equation of the form $\sin(\text{angle}) = \text{number}$.
- Since the interval of allowable values of x is $0 \le x \le 90$ only the 1st quadrant value is considered in this particular example.

Solve for x
- $x - 18·4... = 18·4... \Rightarrow x = 18·4... + 18·4...$

HMRN: p 54

2.

Use $\cos\theta = \dfrac{v.w}{|v||w|}$

when $v = \begin{pmatrix} -2 \\ 3 \\ 5 \end{pmatrix}$ and $w = \begin{pmatrix} 1 \\ -1 \\ 3 \end{pmatrix}$ ✓

$v.w = \begin{pmatrix} -2 \\ 3 \\ 5 \end{pmatrix} \cdot \begin{pmatrix} 1 \\ -1 \\ 3 \end{pmatrix}$

$\qquad = -2 \times 1 + 3 \times (-1) + 5 \times 3 = 10$ ✓

$|v| = \sqrt{(-2)^2 + 3^2 + 5^2} = \sqrt{4+9+25}$

$\qquad = \sqrt{38}$ ✓

$|w| = \sqrt{1^2 + (-1)^2 + 3^2} = \sqrt{1+1+9}$

$\qquad = \sqrt{11}$ ✓

so $\cos\theta = \dfrac{10}{\sqrt{38}\ \sqrt{11}}$

$\Rightarrow \theta = \cos^{-1}\left(\dfrac{10}{\sqrt{38}\ \sqrt{11}}\right)$ ✓

$\Rightarrow \theta = 60·71... \doteq 60·7$ (to 1 dec. place)

5 marks

Strategy
- The formula given to you in the exam is: "$a.b = |a||b|\cos\theta$, where θ is the angle between a and b" so you must know to rearrange this into the form used in this question: $\cos\theta = \dfrac{a.b}{|a||b|}$

  As is the case for all formulae: LEARN THEM!

Dot product
- There is a check you can do with the 'dot product'

  If $a.b > 0$ the angle is acute $(0° < \theta° < 90°)$

  If $a.b < 0$ then the angle is obtuse $(90° < \theta° < 180°)$

Magnitudes
- You will be awarded 1 mark for each correct answer.

Angle
- Many mistakes are made in this calculation. Use:

  $\boxed{\text{INV}}\,\boxed{\cos}\,\boxed{(}\boxed{1}\boxed{0}\,\boxed{\div}\,\boxed{(}\boxed{\sqrt{}}\,\boxed{3}\boxed{8}\,\boxed{\times}\,\boxed{\sqrt{}}\,\boxed{1}\boxed{1}\boxed{)}\boxed{)}$ as the keying sequence followed by $\boxed{\text{EXE}}$ (true for most calculators).

  HMRN: p 46

Strategy
- The positive/zero/negative nature of the discriminant of this quadratic equation has to be determined.

3.

$x^2 - 2x + c^2 + 2 = 0$ ✓

Discriminant

$= (-2)^2 - 4 \times 1 \times (c^2 + 2)$ ✓

$= 4 - 4(c^2 + 2) = 4 - 4c^2 - 8$

$= -4c^2 - 4 = -4(c^2 + 1)$ ✓

Since $c^2 + 1 > 0$ then $-4(c^2 + 1) < 0$

So for all values of c the

Discriminant < 0 and the

equation has no Real roots. ✓

4 marks

Substitution
- Comparing $ax^2 + bx + c' = 0$

  with $1x^2 - 2x + (c^2 + 2) = 0$

  gives $a = 1$, $b = -2$ and $c' = c^2 + 2$
  (different c's!)

  So $b^2 - 4ac' = (-2)^2 - 4 \times 1 \times (c^2 + 2)$

Simplify
- Be careful with negatives,
  e.g. $-4(c^2 + 2) = -4c^2 - 8$.

Proof
- $-4 \times (c^2 + 1)$ is negative $\times$ positive = negative
  $c^2 + 1$ is always positive since c^2 is always positive or zero.

HMRN: p 27

4. (a)

$$y = \frac{1}{3}x^3 - 2x^2$$ ✓

$$\Rightarrow \frac{dy}{dx} = x^2 - 4x$$ ✓

The tangent has gradient -4

so $\dfrac{dy}{dx} = -4$ ✓

$\Rightarrow x^2 - 4x = -4$

$\Rightarrow x^2 - 4x + 4 = 0$ ✓

$\Rightarrow (x - 2)(x - 2) = 0$

$\Rightarrow x = 2$ which is the required
  x-coordinate. ✓

5 marks

Strategy
- Knowing to differentiate will earn you this mark.

Differentiate
- Notice the coefficients: $3 \times \frac{1}{3} = 1$ and $2 \times (-2) = -4$.

Strategy
- This 2nd strategy mark is for setting the gradient equal to -4.

Solve
- 2 marks are available: for starting the process and then completing it. You should recognise a quadratic equation and write it in 'standard form', i.e. $x^2 - 4x + 4 = 0$.

HMRN: p 20

4. (*b*)

When $x = 2$

$$y = \frac{1}{3} \times 2^3 - 2 \times 2^2 = \frac{8}{3} - 8 = -\frac{16}{3} \quad \checkmark$$

A point on the tangent is

$$\left(2, -\frac{16}{3}\right)$$

and the gradient $= -4$

So the equation is:

$$y - \left(-\frac{16}{3}\right) = -4(x - 2)$$

$$\Rightarrow y + \frac{16}{3} = -4x + 8$$

$$\Rightarrow 3y + 16 = -12x + 24$$

$$\Rightarrow 3y + 12x = 8 \quad \checkmark$$

2 marks

Calculation
- To find the equation of the tangent you will need to know the coordinates of a point on that tangent. All you know so far is the x-coordinate ($x = 2$) and so calculation of the y-coordinate is essential.

Equation
- Here you are using $y - b = m(x - a)$ where $m = -4$ and the point (a, b) is $\left(2, -\frac{16}{3}\right)$

- Do not use decimal approximations in coordinate work - you will lose marks if you do, e.g. $-5\cdot3$ should not be used for $-\frac{16}{3}$.

HMRN: p 20

5. (*a*)

$C_t = C_0 e^{-\frac{1}{4}}$

In this case $C_t = 3\cdot5$ when $t = 3$

$$\Rightarrow 3\cdot5 = C_0 e^{-\frac{3}{4}} \Rightarrow 3\cdot5 = \frac{C_0}{e^{\frac{3}{4}}} \quad \checkmark$$

$$\Rightarrow C_0 = 3\cdot5 \, e^{\frac{3}{4}} = 7\cdot409\ldots \quad \checkmark$$

The concentration just after administration was $7\cdot4$ mg/ml $\checkmark$ (to 1 decimal place).

3 marks

Strategy
- In a question like this you should know that you will have to pick out values for the letters in the formula and do a substitution. Sometimes it helps to label the formula:

$$C_t = C_0 \, e^{-\frac{t}{4}}$$

concentration after t hours · concentration at the start · the time elapsed (t hours)

Change of Subject
- C_t is the subject of the given formula. You are asked to find C_0 - the concentration initially. This means you aim for $C_0 = $ (a number)

Calculation
- On the calculator use.

$$\boxed{3}\boxed{\cdot}\boxed{5}\boxed{\times}\boxed{e^x}\boxed{(}\boxed{3}\boxed{\div}\boxed{4}\boxed{)}\boxed{\text{EXE}}$$

HMRN: p 50–51

5. (b)
Required to find t so that

$$C_t = \frac{1}{2}C_0$$

$$\Rightarrow \frac{1}{2}C_0 = C_0 e^{-t/4} \quad \checkmark$$

$$\Rightarrow \frac{1}{2} = e^{-t/4} \quad \checkmark$$

$$\checkmark$$

$$\Rightarrow \log_e \frac{1}{2} = -\frac{t}{4}$$

$$\Rightarrow t = -4\log_e \frac{1}{2} = 2\cdot 772\ldots$$

This is 2 hours and $0\cdot 772\ldots \times 60$ min.
It takes 2 hours 46 minutes $\quad \checkmark$
(to the nearest minute).

4 marks

Interpretation
- You are being asked to calculate t for C_t, the concentration after t hours, to equal half of the initial concentration, i.e. $\frac{1}{2}C_0$.

1st step to solving
- Both sides of the equation can be divided by C_0. This creates an exponential equation with only the variable t.

Log Statement
- You use this conversion:
$c = b^a \leftrightarrow \log_b c = a$

In this case $a = -\frac{t}{4}$, $b = e$ and $c = \frac{1}{2}$

Calculation
- The $\boxed{\ln}$ key calculates '$\log_e$'.

HMRN: p 50–51

6. (a)
$x^2 + y^2 - 4x - 6y + 8 = 0$
Centre: $C_1(2, 3)$ $\quad \checkmark$
For $C_1(2,3)$ and $A(1,5)$ $\quad \checkmark$

$$m_{C_1 A} = \frac{5-3}{1-2} = \frac{2}{-1} = -2$$

$$\Rightarrow m_\perp = \frac{1}{2} \quad \checkmark$$

Point on the tangent is $A(1,5)$

and the gradient $= \frac{1}{2}$
so the equation is:

$$y - 5 = \frac{1}{2}(x-1)$$

$$\Rightarrow 2y - 10 = x - 1$$

$$\Rightarrow 2y - x = 9 \quad \checkmark$$

4 marks

Centre
- This uses the result:

circle: $x^2 + y^2 + 2gx + 2fy + c = 0$

Centre: $(-g, -f)$

The process involves halving and changing signs of the coefficients of x and y to get the coordinates of the centre of the circle.

Gradient of radius
- The gradient formula is:

$$P(x_1, y_1), Q(x_2, y_2) \Rightarrow m_{PQ} = \frac{y_2 - y_1}{x_2 - x_1}$$

Strategy
- The tangent is perpendicular to the radius to the point of contact, i.e. $C_1 P$

- Perpendicular gradients are obtained by using $m = \frac{a}{b} \Rightarrow m_\perp = -\frac{b}{a}$. In this case -2 is thought of as $-\frac{2}{1}$.

Equation
- Use $y - b = m(x - a)$. In this case you use $m = \frac{1}{2}$ with (a, b) being the point $A(1,5)$.

HMRN: p 3–4, p 39

6. (*b*)

For intersection of line and circle solve:

$$2y - x = 9$$
$$x^2 + y^2 + 2x + 2y - 18 = 0 \Big\} \quad ✓$$

Substitute $x = 2y - 9$ in circle equation: ✓

$(2y - 9)^2 + y^2 + 2(2y - 9) + 2y - 18 = 0$

$\Rightarrow 4y^2 - 36y + 81 + y^2 + 4y - 18$
$\quad + 2y - 18 = 0$

$\Rightarrow 5y^2 - 30y + 45 = 0 \quad ✓$

$\Rightarrow 5(y^2 - 6y + 9) = 0$

$\Rightarrow 5(y - 3)(y - 3) = 0$

$\Rightarrow y = 3 \quad ✓$

and since there is only one solution the line is a tangent to the circle. ✓

5 marks

Rearrangment
- Change $2y - x = 9$ to $x = 2y - 9$ ready for substitution.

Strategy
- The strategy mark here is awarded for substitution of the line equation into the circle equation.

'Standard form'
- You should recognise a quadratic equation and therefore write it in the standard order, namely:

$5y^2 - 30y + 45 = 0$

Solve
- It is always easier to factorise quadratic expressions if any common factor is removed first. In this case 5 is the common factor.

Justify
- How do you prove a line is tangent to a circle? You find the points of intersection. If the line is a tangent there will be only 1 point. You write a clear statement to this effect to gain this 'communication' work.

HMRN: p 40

---

6. (*c*)

When $y = 3$
$x = 2 \times 3 - 9 = -3$ ✓
so the point of contact is B$(-3, 3)$
For A$(1, 5)$ and B$(-3, 3)$

$AB = \sqrt{(1 - (-3))^2 + (5 - 3)^2}$

$\quad = \sqrt{4^2 + 2^2} = \sqrt{20} = 2\sqrt{5}$ ✓

2 marks

Other coordinate
- Having determined the value of the y-coordinate in part (*b*) you now have to substitute this back (use the line equation) to find the x-coordinate.

Distance
- You use the distance formula: P(x_1, y_1), Q(x_2, y_2)

$PQ = \sqrt{(x_2 - x_1)^2 + (y_2 - y_1)^2}$

HMRN: p 7

7. ✓

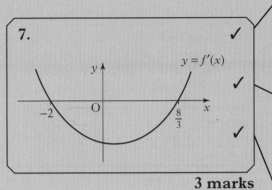

✓

✓

3 marks

Stationary points
- Any stationary point on the graph $y = f(x)$ has a zero value for the gradient. So on $y = f'(x)$, the graph that shows the gradient values, an x-axis intercept (zero value) indicates a stationary point on the original graph. In this case your sketch should cross the x-axis at $(-2, 0)$ and $(\frac{8}{3}, 0)$.

Correct relationship
- Between -2 and $\frac{8}{3}$ the original graph goes downhill

$\Rightarrow$ gradient is negative $\Rightarrow y = f'(x)$ graph is below the x-axis

- Less than -2 and greater than $\frac{8}{3}$ original graph goes uphill $\Rightarrow$ gradient positive $\Rightarrow y = f'(x)$ graph is above the x-axis

Shape
- Differentiating a cubic produces a quadratic $\Rightarrow$ your graph should show a parabola.

HMRN: p 19

8. (*a*)
Volume $= x \times x \times h = x^2 h$ ✓

so $x^2 h = 62 \frac{1}{2} \Rightarrow h = \dfrac{62\frac{1}{2}}{x^2} = \dfrac{125}{2x^2}$

Base area $= x^2$
Side area $= xh$

$= x \times \dfrac{125}{2x^2} = \dfrac{125}{2x}$ ✓

Total area $= 4 \times$ side area + Base area

$\Rightarrow A(x) = 4 \times \dfrac{125}{2x} + x^2 = \dfrac{250}{x} + x^2$ ✓

3 marks

Strategy
- Volume = length × breadth × height.

Strategy
- Finding h in terms of x $\left(h = \frac{125}{2x^2}\right)$ and then substituting in the area expressions is the key to solving this question.

Proof
- Make sure all steps are shown. The answer is given so the examiner needs to see each step of your reasoning.

8. (b)

$A(x) = 250x^{-1} + x^2$ ✓

$\Rightarrow A'(x) = -250x^{-2} + 2x$

$\qquad = -\dfrac{250}{x^2} + 2x$ ✓

For stationary values set

$A'(x) = 0$ ✓

$\Rightarrow \dfrac{-250}{x^2} + 2x = 0 \;(\times x^2)$

$\Rightarrow -250 + 2x^3 = 0 \Rightarrow 2x^3 = 250$

$\Rightarrow x^3 = 125 \Rightarrow x = 5$ ✓

$$x: \qquad \underrightarrow{\quad\;\; \overset{5}{\bullet} \quad\quad}$$

$A'(x) = -\dfrac{250}{x^2} + 2x: \qquad - \quad \{ \quad +$

Shape of graph: $\qquad \diagdown \, - \diagup$

nature: $\qquad$ min ✓

So $x = 5$ gives a minimum value
for the area.

5 marks

Preparation
- You should know that the term $\frac{250}{x}$ cannot be directly differentiated. Changing the expression to $250x^{-1}$ fits it to the following rule:

  $f(x) = ax^n \Rightarrow f'(x) = nax^{n-1}$

  in this case $a = 250$ and $n = -1$.

Differentiation
- It is useful for subsequent work to change negative indices to positive. In this case $-250x^{-2} = -\frac{250}{x^2}$

Strategy
- '$A'(x) = 0$' has to be stated to gain this mark.

Solve
- You should multiply both sides by x^2. The aim is to rid the equation of fractions.

- Cubing a positive number gives a positive, but cubing a negative number gives a negative, so $x^3 = 125$ has only one positive solution. This is not like squaring: $x^2 = 25$ gives $x = 5$ or $x = -5$.

Justify
- 1 mark is allocated for the 'nature table'.

  HMRN: p 20–22

9.

First find the points of intersection of $y = 5$ with the curve:

Solve:

$$\left. \begin{array}{l} y = 5 \\ y = x^2 - 6x + 10 \end{array} \right\} \quad \begin{array}{l} x^2 - 6x + 10 = 5 \\ \Rightarrow x^2 - 6x + 5 = 0 \\ \Rightarrow (x-1)(x-5) = 0 \\ \Rightarrow x = 1 \text{ or } x = 5 \end{array}$$

✓
✓

Diagram:

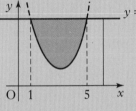

shaded area in this diagram is given by: $\int_1^5 5 - (x^2 - 6x + 10)\, dx$ ✓

$$= \int_1^5 5 - x^2 + 6x - 10\, dx$$ ✓

$$= \int_1^5 -x^2 + 6x - 5\, dx$$ ✓

$$= \left[-\frac{x^3}{3} + 3x^2 - 5x \right]_1^5$$

$$= \left(-\frac{5^3}{3} + 3 \times 5^2 - 5 \times 5 \right) - \left(-\frac{1^3}{3} + 3 \times 1^2 - 5 \times 1 \right)$$

$$= -\frac{125}{3} + 75 - 25 + \frac{1}{3} - 3 + 5$$

$$= 52 - \frac{124}{3} = \frac{156}{3} - \frac{124}{3} = \frac{32}{3} \text{ cm}^2$$ ✓

So Area of plate $= 5 \times 6 \qquad -\frac{32}{3}$

(area of rectangle)

$$= 30 - \frac{32}{3} = \frac{90}{3} - \frac{32}{3} = \frac{58}{3}$$

$$= 19\frac{1}{3} \text{ cm}^2$$ ✓

8 marks

Strategy
- The strategy is to find the intersection points of line $y = 5$ and parabola $y = x^2 - 6x + 10$ as the x-values of these points will subsequently be crucial in setting up integrals to find areas.

Solve
- Always rearrange a quadratic equation into 'standard form', in this case $x^2 - 6x + 5 = 0$ so that the factorisation allows this process: (factor1)(factor2) = $0 \Rightarrow$ factor1 = 0 or factor2 = 0

Strategy
- This strategy mark is for how you will split areas up. In the solution given the approach is:

  (Rectangle Area) – (Parabolic Area).

  An alternative approach would have been to add together 3 areas to calculate the plate area as shown in this diagram. area A + area B + area C

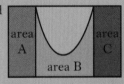

Strategy
- Did you know to integrate to find this area?

Limits
- Left to right on the diagram (1 and 5).

  Bottom to top on the integral (1 and 5).

Integration
- You should always 'simplify' first before you integrate. In this case don't integrate $5 - (x^2 - 6x + 10)$, simplify first to $-x^2 + 6x - 5$ and then integrate.

Evaluate
- Even with a calculator available these calculations are very difficult to get correct. You should take great care and double-check your working always.

Area
- Finally make sure you answer the question. $\frac{32}{3}$ cm^2 is not the required answer! Your strategy was to subtract the parabolic area from the rectangle.

HMRN: p 32–33